MATHEMATICS FOR TH

LUCK AND JUDGEMENT

Other Books in the Series

*

MATHEMATICAL EXPERIENCE
MACHINES, MECHANISMS AND MATHEMATICS
ASSIGNMENT SYSTEMS
MATHEMATICAL PATTERN
NUMBER APPRECIATION

MATHEMATICS FOR THE MAJORITY

LUCK AND JUDGEMENT

a classroom approach to probability and statistics

CHATTO & WINDUS
LONDON
FOR
The Schools Council
1971

Published by
Chatto & Windus (Educational) Ltd
42 William IV Street
London W.C.2
*
Clarke, Irwin & Co. Ltd
Toronto

ISBN 0 7010 0464 9

Printed by photo-lithography and
made in Great Britain at
the Pitman Press, Bath

Contents

Mathematics for the Majority

The Schools Council Project in Secondary School Mathematics (now called *Mathematics for the Majority*) was set up to help teachers construct courses for pupils of average and below-average ability that relate mathematics to their experience and provide them with some insight into the processes that lie behind the use of mathematics as the language of science and as a source of interest in everyday things.

Members of the Project Team (1969–71):

P. J. Floyd (Director)
J. H. D. Parker (Deputy Director)
K. C. Bonnaud
T. M. Murray Rust
E. T. Norris
Mrs J. Stephens
P. Kaner (Evaluator)

The chief author of this book is:

Mrs J. Stephens

Introduction

The topics of Probability and Statistics are a popular choice for all secondary school pupils. This is because, apart from the scientific applications, one has only to open a newspaper or watch television to be confronted by data of all sorts, graphical and other interpretations of them, forecasts, insurance, advertisers' claims and opinion polls. There is no doubt, therefore, that it is desirable that all our pupils should have the opportunity to know some of the ways in which facts can be well – and badly – collected and used for various ends: for it is certain that even those who are not going to be involved in doing this themselves will not be able to escape receiving the products of such activity.

We consider that a practical approach, rather than a theoretical one, is more appropriate within our frame of reference. Hence the work suggested in this book centres around classroom experiments for pupils to undertake singly or in small groups. *The activities are aimed at providing the pupil with a practical basis for the classroom discussions, which are themselves the main vehicle by which he may come to appreciate some of the ideas of probability and statistics, especially as they are likely to affect him.*

It is because our approach develops from a series of practical experiences that you will not find here a supply of theoretical examples to be worked. Those who feel that such examples are indispensable will find plenty of them in some of the books listed in Appendix 1. *Our* hope is to develop in our pupils a critical and informed approach based partly on their everyday experiences and partly on work arising from planned experiments. We do *not* aim to produce pupils whose knowledge of the topics consists solely of the application of standard mathematical techniques to standard problems.

We do not recommend that all the work suggested here should be presented continuously from beginning to end. It could best be spread throughout the secondary school years and many teachers will find that some groundwork has been done in the primary school. It is recognized also that some teachers who will be using this material will not themselves have a very strong mathematical background in these topics. For this reason possible outcomes of experiments and pointers for discussion are given in a detail which the mathematically qualified may feel to be unnecessary. We hope that teachers who use this book will adapt or reject the ideas put forward according to their *own* needs and experience, as well as that of their pupils, and that the material will be a growing point and not an end.

PART 1

Chapters 1-5

We begin with some suggestions for introductory discussions and an indication of the ideas which might arise from those discussions. There follows, in Chapter 2, a sample of activities for groups or individuals in the classroom and it is from these activities that we develop the rest of the work.

Chapter 3 is a digression on the subject of 'graphs' and is inserted at this point simply to assist teachers who are not very familiar with the two main types of graph used in schools. Knowledge of these types is assumed in Chapters 4 and 5.

These last two chapters form the bulk of Part 1 and give full commentary on the activities of Chapter 2 including sample results, suggestions for tabulation and graphing, analysis of some of the mathematical content and general pedagogy.

1

Preliminary Group Discussion

Intuitive ideas in everyday life – suggestions for group discussion and its outcome – types of probability, i.e. situations involving past records and the assumptions made; situations involving current statistics and sampling and assumptions made in sampling; *a priori* probability involving finite outcomes of events and the equal likelihood assumption.

We all build up, through our everyday experiences, some sort of idea of the likelihood or otherwise of the occurrence of particular events.

'It looks as though it will rain.'

'Arsenal will beat Aston Villa.'

'We might finish this before tea.'

'I expect so.'

'It's an even chance that . . . '

'It's more than likely that . . . '

Such are some of the remarks which show our intuitive knowledge of *probability*. Thoughtful listening to conversation, T.V. and radio programmes may produce many more examples. One notices that most of the comments are based on a knowledge of the past, on a crude analysis of previous events and an *assumption* that things will continue in the same way. The last two phrases in the list, however, indicate some idea of comparing probabilities (the first an equality, the second an inequality) and this points to the possibility of some numerical *measurement* of probability.

In introducing the idea of probability to a group of children the teacher has, therefore, a pool of their past experiences from which the mathematical idea may grow. Discussion of a few situations with the whole class or with a group will set the scene for experiment and data collection later. The situations suggested fall into two broad categories, those dependent on statistical records and those where all possibilities can be examined in advance. These categories are fully discussed later in the chapter.

(a) If a 100 people are chosen at random is it likely that at least one of them will be called Smith?

(b) If you buy a ticket in a raffle what is your chance of winning?

(c) If you drop a slice of buttered toast on to the floor how likely is it to fall butter side down?

(d) If I toss a fair coin 4 times am I likely to get at least one head? What is the likelihood if I only toss it once? Does the likelihood change for 2, 3 or 5 or 6 tosses?

(e) Is it likely that at least two people in the class were born in the same month?
(f) Is it likely that at least two people in the class have the same birthday? Is the likelihood the same for a class of a different size? (The birthday problem, which is well known and has, for most people, a rather surprising result, is discussed further in Chapter 4.)
(g) Why do car insurance brokers charge higher rates for drivers under 25 years old?
(h) How does the Meteorological Office make long range weather forecasts?
(i) How likely is it that the football (hockey) team will win its next match?
(j) For what proportion of throws would you expect to get a six if you repeatedly threw a fair cubical die?
(k) If you choose one card from a pack of playing cards is it likely to be a red card?
(l) Why do sportsmen toss a penny to decide who starts a match?
(m) How does insurance work? What are the likely results of premiums being too low or too high?
(n) How likely is a White Christmas?
(o) How likely is it that the next Bank Holiday will be wet?
(p) How likely is it that you will win at 'Roll-a-penny' at a fairground?
(q) How likely is it that a man will live to be 80 years old?
(r) If we choose one person in this room at random how likely is it that he or she has one brother?

We do not expect specific numerical or even very full answers at this stage (although one might well return to such questions later expecting more logical replies – some would make useful assessment questions). Some pupils make very wild and subjective judgements at first. Others rightly say to most of the situations, 'It depends'. The discussion then moves to consider on *what* it depends. It is through consideration of this point that pupils can come to realize what the problems are and to see what information would be needed to arrive at some conclusion.

By the end of the discussion some classification of the situations and their probabilities should have begun to emerge as follows. The only way of making reasoned statements about the probabilities in some of the situations, for example: (g), (h), (m), (n), (o) and (q), is by having detailed records of past events *and* assuming that future events will follow the pattern, if any, of the past. *This assumption is basic and pupils need to be made specifically aware of it if they are to understand this method of assessing probability.* If we do not make this assumption there is clearly no point in consulting records and any judgement of probability we make is entirely subjective. Discussion should include some consideration of just what records would be needed and by whom they are usually kept (if at all).

Other situations, for example (a) and (r), depend not on past statistics but on a knowledge of statistics of the present. In some cases like this it is virtually impossible to obtain accurate statistics for the whole population under consideration. 'Population' in this context refers not necessarily to people but represents everything about which information is being collected.

If a study of the entire population is impracticable then one of the problems is to know what proportion must be examined (sampled) in order to make it reasonable to assume that the whole population behaves in the same way. Another problem is the matching of the sample with the population to try to ensure that the sample is really representative of it. The theory of sampling, with its consequences for commerce, industry, science and administration, is dependent on such assumptions.

Some of the situations listed (e.g. (b), (c), (d), (e), (f), (j), (k)) do not seem to be dependent on statistical information but on the finite (i.e. countable) number of possible outcomes of the events. In (j), for example, most pupils will say that they would expect to throw a six for one-sixth of the throws. Their childhood experience of dice games will preclude most from expecting one six in *every* six throws. Without any formal knowledge of the ways in which we calculate probability such pupils are realizing that there are six possible results and they are *assuming* (unconsciously) that all six are *equally likely*. Such an assumption would not be valid in the case of a loaded die, for example. The assigning of a measure of probability before an event takes place, by consideration of all the possible outcomes, is known as *a priori* probability. The equal likelihood assumption, and it *is* only an assumption as we cannot prove it, is basic to this type of probability. Is it reasonable to assume that the two possibilities in (c) (the buttered toast) are equally likely? Is the next child born in the family equally likely to be a boy or a girl? Is a fair penny as likely to fall HEADS as TAILS? Is a dropped drawing pin as likely to fall head down as sideways? We cannot assume that if an event has only two outcomes they are both equally likely. The sum of scores on two tossed dice can be anything from 2 to 12 – are all these equally likely? (If you think so then toss two dice about 100 times – you *may* feel less convinced!) The question of equal likelihood arises again in several of the activities suggested later in Chapter 2.

In summary then we have, broadly speaking, two ideas. Firstly, probability based on past or present statistics (with an assumption about future events following the same pattern, if any) and, secondly, probability in situations where all possible outcomes can be examined beforehand (with an assumption about which outcomes are equally likely). These two major ideas will be met again and again in the activities to follow and it is through teacher/pupil discussion following the activities that we would expect the full understanding of them to develop – the preliminary discussions suggested in this Chapter just serve to bring them to consciousness.

2

Classroom Activities

Methods of working and presentation of material – list of suggested activities for pupils.

After the introductory discussions (see Chapter 1) pupils work individually, in pairs, or in small groups, at practical assignments which are followed by analysis and discussion. The particular environment, the ability of the pupils, the availability of time and apparatus are factors which affect the range of activities which any one teacher provides and the number from the range that any one pupil attempts. It is not, of course, necessary for all pupils to be working at probability and statistics. It is important, however, that each chosen activity is studied by at least three groups in cases when the results are to be compared.

Our choice of activities has been influenced by their suitability for group or individual work, their feasibility from the organizational point of view in most secondary schools, their emphasis on data collected *by* pupils, the variety of ideas that they use or can lead to and the simplicity and availability of apparatus.

A longer list of activities than most classes would need is given and more may be found by consulting some of the books listed in Appendix 1. It may be appropriate to reserve some activities for use in an assessment situation if one is considered necessary or desirable. Most teachers find that the pupils we are considering cannot cope with detailed written instructions for procedure *and* recording *and* presentation *and* interpretation of results. *We suggest, therefore, that, where possible, pupils are given only such brief written instructions as will enable them to proceed with the activity and that detail of presentation and interpretation comes by discussion, both during and after the activity, thus giving each pupil the opportunity to make his own contribution.* Where ideas are common to several activities larger groups can be included in discussion and as pupils absorb more ideas and methods they should need less help (both from written instructions and from the teacher) than they did at first. Suggestions regarding pupils' written recording are made in Appendix 3.

The activities are all listed together in this present chapter, and are also reprinted in Chapters 4 and 5, which contain notes on presentation, interpretation, and follow-up. In this way we hope first to give an overall picture of the scope of the work, and then, in the notes, to draw together several activities which have features in common. The numbering of the activities is for easy reference and implies no order of working. The page reference immediately below each card indicates where the follow-up notes are to be found.

1. If you tossed four coins 50 times about how many times do you *think* you would get at least two HEADS in the four tosses? Try it. Then repeat the experiment using just two coins, then three coins. About how often do you think you would get at least two HEADS if you tossed *five* coins 50 times?

(pp.27, 28)

2. Do you think that it is likely that two people in the class have the same birthday? Find out. Get information about this from other classes as well.

(pp.59, 61)

3. Do you think that out of five people two at least are likely to have the same birth month? Write down the birth months of fifty sets of five people. In how many of your sets were at least two people born in the same month?

(p.59)

4. If you throw four dice fifty times how often do you think you will get at least one SIX? Try it. Then try it with just one die, then with two dice, then with three. About how often do you think you would throw at least one SIX if you threw five dice fifty times?

(pp.36, 38)

5. If you have five playing cards including just one ACE and you choose one card, and then replace it, fifty times, how many times do you think you will choose the ACE? Try it. Then repeat the experiment but draw two cards at a time, then draw three cards at a time. About how often do you think you would get the ACE if you drew four cards at a time? How often if you drew five cards?

(p.51)

6. If you roll a coin or disc 100 times down a chute, made from a piece of card folded, onto a board marked off in squares whose sides are twice the length of the diameter of the coin, how often would you expect the coin to land completely inside the square? Try it. If you were to double the length of the side of the square how often would you expect it then? Suppose you doubled the diameter of the coin instead?

(pp.63, 64)

7. If you toss five coins many times what number of HEADS out of five do you think is likely to appear most often? What number would you expect least? Do the experiment and record the number of HEADS each time. Can you think of a reason for the pattern of the results?

(pp.27, 32)

8. If you toss two dice many times and add together the two scores each time, which total score do you think you will get most often? What are the highest and lowest total scores possible? Which score do you think will turn up least often? Do the experiment and record all the scores. Can you think of an explanation for the results?

(pp.36, 41)

9. If you throw three dice together and add the scores, what are the lowest and highest totals you can possibly get? Which scores would you expect to turn up most often and least often? Do the experiment by throwing three dice at least 300 times. Can you think of an explanation of the results?

(pp.36, 41)

10. If you throw three dice together do you think it is most likely that you will get three SIXES or two SIXES or one SIX or no SIXES? Throw the three dice about 300 times and record the number of sixes each time.

(pp.37, 38)

11. There are 100 cubes in the box and each has one marked face. If you throw them onto the table how many do you think will land marked side uppermost? Do it and record the number. Then remove the cubes which have fallen marked side up and throw the *rest* down again. Keep on doing this until you have no cubes left. Repeat the experiment from the start several times. About how many cubes do you *expect* to show their marked side after the third throw?

(p. 48)

12. If you throw two dice many times and record the *difference* between the two scores, what is the smallest possible difference? What is the largest possible difference? Which differences do you think will turn up most often and least often? Throw the two dice about 100 times. Are the results as you expected? Can you explain them?

(pp.37, 45)

13. One of these two dice is biased (that means it is an unfair die). Can you find out, by some tossing experiments, which one it is and to which number or numbers it is biased?

(pp.37, 48)

14.

1					F
				E	
			D		
		C			
	B				
A					2

Two players (1) and (2) take 16 counters each and place them in the squares marked 1 and 2 respectively. Then they take it in turns to toss a coin. On each throw, if (1) tosses a HEAD he moves his first counter one place to the right → and if a TAIL he moves it one place down ↓ . When (2) tosses a HEAD he moves his first counter one place to the left ← and for a TAIL one place up ↑ . This goes on until the first counter reaches any of the squares lettered A, B, C, D, E, F. When a counter gets 'home' then that player starts a new counter. The winner is the player who, at the end, has most counters in squares A and F *together.* If this is a tie then the counters in B and E are counted instead. Play the game several times. Is there a pattern in the results?

(pp.27, 35)

15. If you toss two dice many times and *multiply* the two scores together do you think that it is equally likely that the answer you get is exactly divisible by three, or gives remainder 1 when divided by three, or gives remainder 2 when divided by three? Throw the dice at least 100 times and after multiplying the two numbers, divide by three and record the *remainder.* Was your prediction correct? Can you find an explanation for the results?

(pp.37, 45)

16. Take a box containing one bead of one colour, three of another colour and five of a third colour. If without looking you choose one bead at a time and then replace it when you have seen its colour so that there are always nine in the box, which colour do you think you are likely to choose least often? Which do you think will come out most often? How do you think the two numbers will compare? Now do the experiment many times. Were you about right?

(pp.53, 55)

17. Take a box with equal numbers of red and blue beads. Choose one bead at random, record the colour and put it back. Then repeat for a second bead. Do you think you are more likely to draw the same colour bead each time or a different colour? Choose two beads at least 100 times and record the results in columns headed RR, BB, RB. Can you explain the results?

(pp.54, 56)

18. Take a box with equal numbers of red and blue beads. Choose one bead at random, record the colour and replace. Do this at least 50 times. After this choose one, replace and choose another and record the colour sequence RR, RB, BR, or BB – repeat this at least 50 times. Lastly do the experiment with three beads, replacing each one directly you have noted its colour so that there are always the same number of beads left in the box.

(pp.54, 57)

19. Take a box of equal numbers of red, blue and yellow beads. Choose one at random many times, record the colour and replace. Next choose one, replace and choose another and record the colour sequence RR, RY ..... etc. Do this many times. After that repeat the experiment for a run of three beads replacing each bead before choosing another so that there are always the same number of beads to choose from. Can you see and explain any pattern in the results?

(pp.54, 57)

20. Make a row of the twelve boxes which each contain all the letters of the alphabet. Starting from one end of the row take one letter at random from each box until you have two letters the same. Record how many boxes you have drawn from. Replace the letters and repeat the experiment many times. If this were a game in which you could choose how many boxes from which to draw, and the prize for success in getting two letters the same was higher the *fewer* boxes you chose, then how many boxes would *you* ask for?

(pp.59, 62)

21. Toss a coin five times and record the number of HEADS. Toss five more times and record the total number of HEADS out of the ten throws so far. Continue in batches of five throws keeping a 'running total' of the number of HEADS and the number of throws. Try to illustrate the results by a graph.

(pp.28, 35)

22. Should a letter rack have 26 pigeon-holes, one for each letter of the alphabet, or would a different arrangement be better? Use a telephone directory to help you to decide on the best arrangement.

(p.67)

23. Should a word building game or a child's printing set have equal numbers of each letter of the alphabet? Use a book or a newspaper to estimate the frequency with which various letters of the alphabet appear in print.

(p.67)

24. Collect information about the number of brothers and sisters which people in this class have. If you choose someone from the class at random, is he or she likely to have at least one brother or at least one sister?

(p.69)

25. Should shoe shops stock equal numbers of all sizes of shoes so that they can be fairly sure of having something to fit any customer? Record the shoe size and foot length (without shoes) of as many people as possible. Is there any relationship between these two measures?

(p.73)

26. Collect as much information as possible about people's foot-length and foot-width (without shoes). Why would information of this type be important to a shoe manufacturer?

(pp.73, 76)

27. Are boys taller and heavier than girls of the same age? Is it likely to be true that for boys armspan is usually more than height and for girls armspan and height are usually the same? Measure the height (without shoes), the weight and the armspan of as many people about your age as possible.

(pp.73, 77)

28. If you tip a lot of drawing pins onto a flat surface what proportion do you think would land head down? Try it by tipping out a full box of drawing pins. Were you about right? Now take just six drawing pins and throw them down repeatedly, recording the number which land head down each time. Is there any relationship between these results and the tipping of the whole box?

(p.84)

29. Do records of attendance in the school or class really support the claim that Monday and Friday are likely to have the highest absentee rate? Can you think of any reasons for the pattern that absences seem to take?

(p.81)

30. Arrange for everyone in the class to measure the length of the room to the nearest centimetre. Do they all give the same measurement? What do you think the results tell you about the length of the room?

(p.86)

31. On any one Saturday in the season about how many goals are likely to be scored altogether by the 1st, 2nd, 3rd and 4th divisions of the Football League? What is likely to be the commonest number of goals scored by any one team? Are more goals likely to be scored by 4th division teams than by 1st division teams? Use past results to help you to find reasonable answers to these questions.

(p.72)

32.

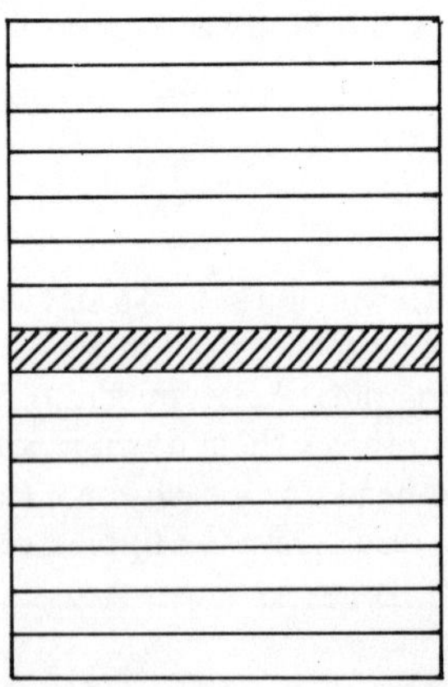

If you knock a coin up the board with a sharp blow of the palm of your hand how often do you think that you will get the *centre* of the coin to land in the coloured band? Try it about 100 times and record in which band you land each time. If you played this with a partner how would you decide who was the best shot?

(p.87)

3

Notes on Graphs

The purpose of a graph – organization of data – development of frequency graphs – graphs of relation between two sets – the question of joining successive points – choice of axes – summary.

(*Note:* This Chapter is a digression from the main course of this book. It is inserted at this point simply to assist teachers who have not done much graph work before, as a knowledge of the two basic types used in schools is assumed in Chapters 4 and 5.

Those who are very experienced in the teaching of graphs, and are knowledgeable about methods used in a wider sphere, may well feel that we have over-simplified the situation. Nevertheless we feel that the two types we discuss are sufficient for the purpose in hand.)

Teachers will often want their pupils to illustrate the results of their experiments or data collections by some sort of graph. A good graph is not just a pictorial representation of some data, but should in some way organize or interpret it. To take a very simple, and hence almost trivial, example, which might occur at the earliest stages of graphing classroom information, the following table shows the number of children favouring each colour. The graph beside it is simply a straightforward 'picture' of the information as given in the table.

Colour	No. of children
Red	6
Green	2
Blue	10
Yellow	4
Purple	1

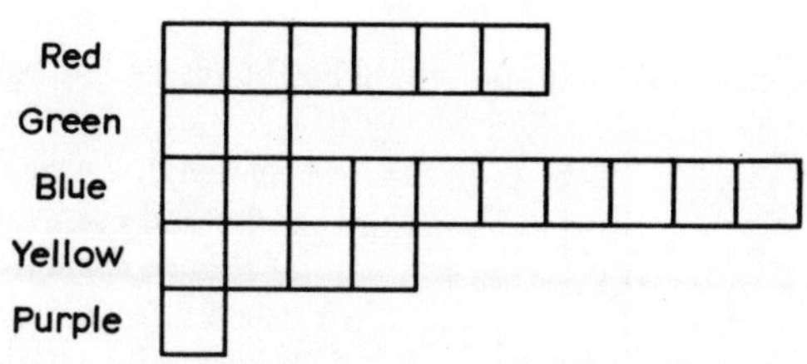

Fig. 1

Clearly there is not a great deal of organizing and interpreting to be done here, but the next graph adds to the table rather than being just an alternative to it simply by *ordering* the data in a useful way, enabling comparisons to be made very easily.

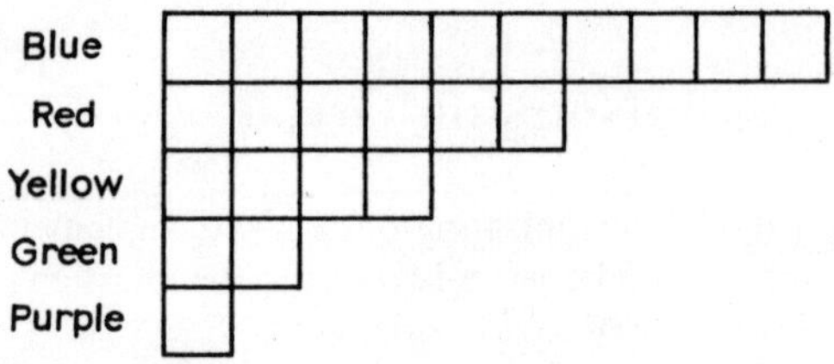

Fig. 2

A second example goes a stage further. The table below (Fig. 3) gives the heights, to the nearest centimetre, of a set of boys.

John	170
Ken	171
Peter	169
Denis	171
Bill	168
David	170
Philip	170
Stephen	170
Richard	171
Jim	170
Fred	169

Fig. 3

As in the last example the data could be put in order and each boy's height represented by a strip. Without drawing this graph it is clear that the strips are going to be very close in length and that nothing extra in the way of interpretation will have been added to the information already given in the table.

A more useful graph would be that given below:

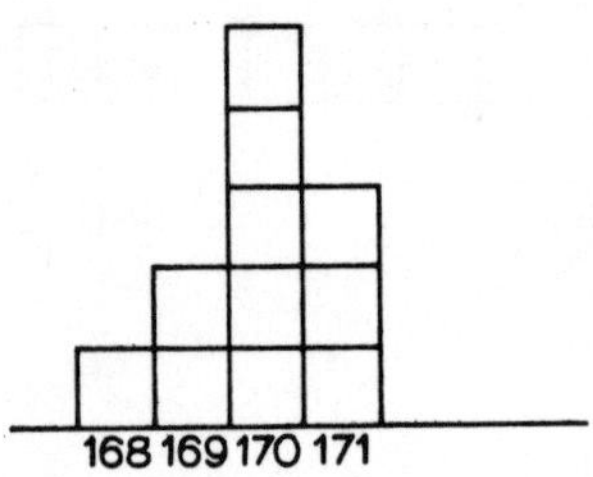

Fig. 4

The heights have been put into ascending order and each square represents one boy of that particular height. Such a graph is called a *frequency* graph because it shows how often something has occurred. This type of graph plays an important part in the work to follow and is one of the two major types of squared-paper graphs in mathematics.

A frequency graph is appropriate whenever data can be split up into separate sets and the number of items in each set can be counted. The very simplest representation of a situation of this type is:

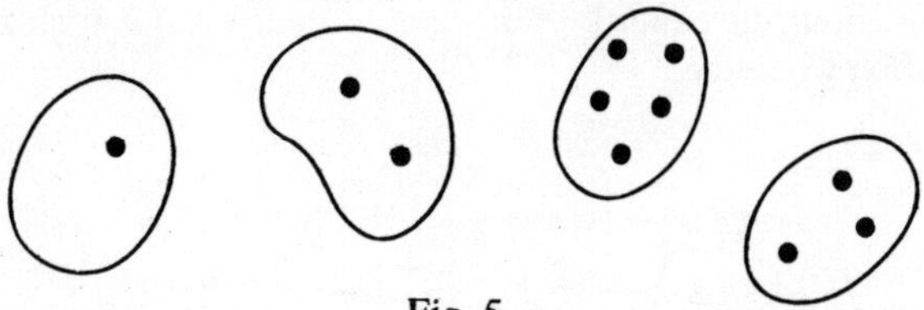

Fig. 5

One can imagine a teacher of very young children arranging twelve skipping ropes as loops on the floor and labelling each with a month. The children would put themselves into the appropriate loop. The resulting 'human' graph would be one like our diagram. (Fig. 5.) Similarly the children may imitate a conventional frequency graph by standing in rows for their birthday months. Later each child may be represented by a matchbox or a cube and finally by a coloured square on graph paper.

Notice that *one* event (i.e. child's birthday month, height, etc.) is represented by *one* unit on the graph. Even with older children, especially the less able, it is advisable to keep to this one-to-one correspondence for some time as the idea of scale is quite a difficult one. Collections of data where the count is so big (e.g. the whole school) that scaling is necessary can be contrived later. We are not suggesting that they should be avoided altogether.

Notice also that the horizontal scale represents the separate sets and that each set takes up a whole section ⊢169⊣ not just a single point —169— . A vertical scale marked with counting numbers is often added for convenience. Many examples of frequency graphs will be found in the notes on the classroom activities.

The second major type of graph used in statistics is one which shows a relationship between individual members from two sets. For example, the table below gives the ages and weights of a sample of a certain British coin.

Age in years	Weight in grammes
15	9·4
45	9·0
50	9·1
30	9·2
30	9·3
25	9·4

Fig. 6

We see that in this particular sample ages 15 and 25 are both related to weight 9·4: similarly age 30 is related to both the weights 9·2 and 9·3 while age 50 is only related to weight 9·1. (We may be justified in drawing some conclusions about the conditions of some of the coins.) A typical diagram of the relation would be:

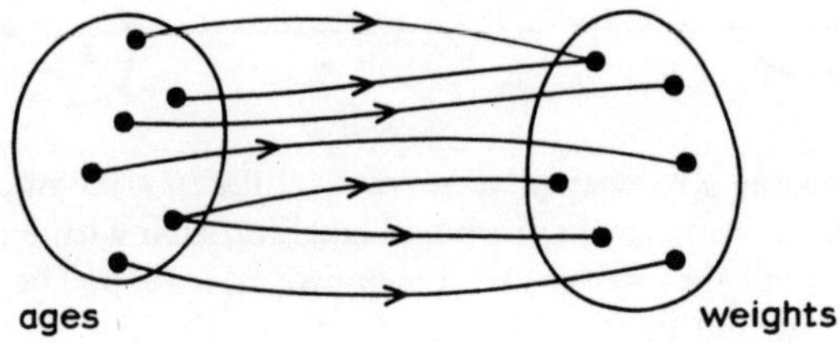

Fig. 7

On squared paper we would have:

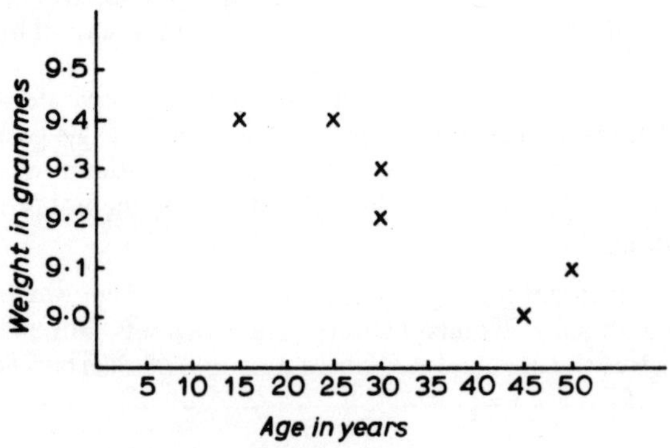

Fig. 8

Notice that we find the position of the x which represents a relationship between one member of each set by travelling at right angles from the appropriate points on the two axes (Fig. 9a), but that it is not usual to mark in either or both of these travelling lines on the graph as it is only the *intersection point* of the two lines which has any significance (Fig. 9b).

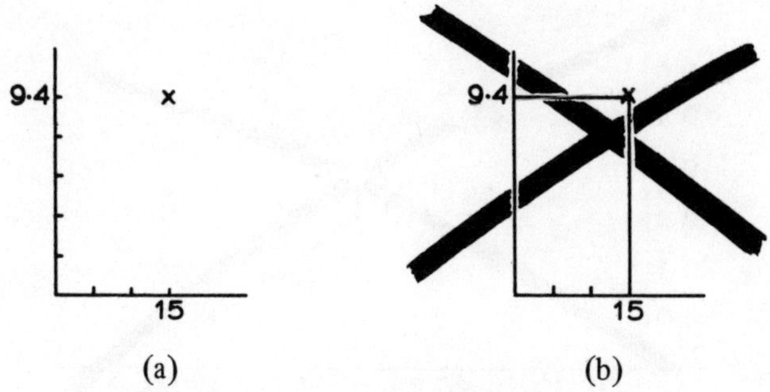

Fig. 9

For the frequency graph, however, we would always (at least at school level) mark in the *whole* column which contributes to the count – we do not mark just the top as this time the whole thing is relevant (Fig. 10).

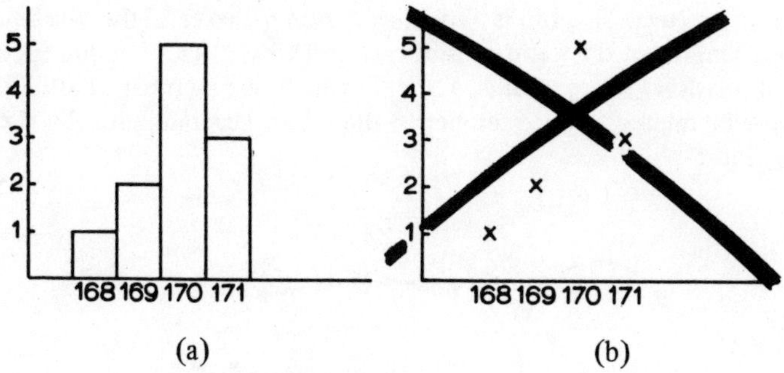

Fig. 10

Notice that in a relation graph involving measurements the axes are marked continuously and just one point (not a whole section as in the frequency graph) represents each particular reading.

A further problem arises in the relation graph and that is whether or not we should join up successive points. If the points *are* joined up then the implication is that all the points lying along the line are relevant to the situation. For example, it is clearly incorrect to join the points in the (coin age) → (weight) graph thus (Fig. 11):

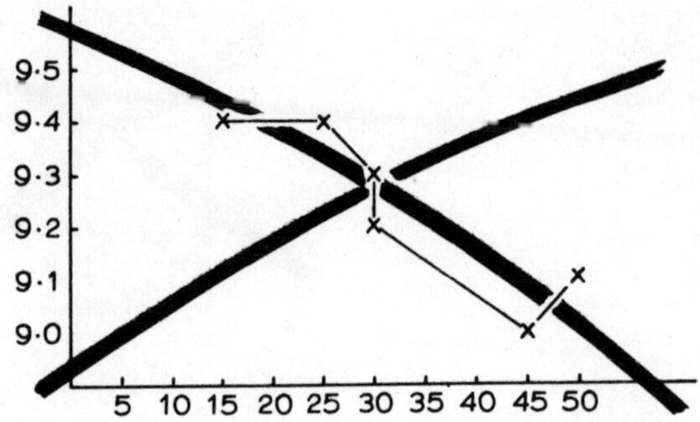

Fig. 11

To do so would imply quite unjustified data about coins not in the sample. Imagine the confusion of lines in this example if the sample had 60 coins instead of 6.

As a counter example, however, consider the graph of (length of side)→(area) for squares. This too is a *relation between two sets:* the sets being possible lengths of sides and possible areas. This is in fact a rather special type of relation known as *one-to-one*, in which one member of either set *can* only be related to one member of the other. Diagrammatically it can appear thus:

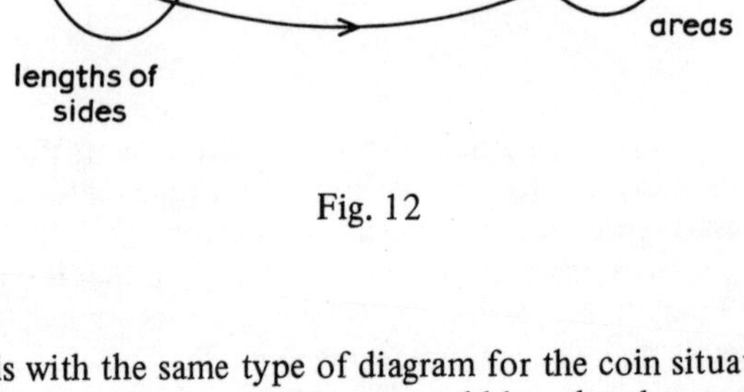

Fig. 12

Compare this with the same type of diagram for the coin situation (Fig. 7), which is not one-to-one as a certain age could be related to more than one weight and vice versa.

As a conventional relation graph we get:

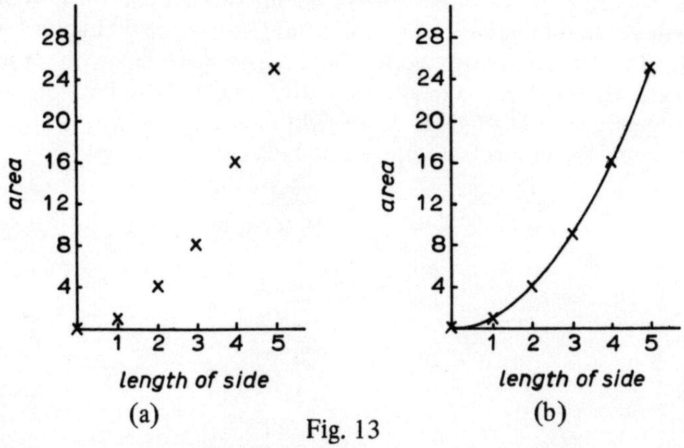

Fig. 13

We know that readings between the marked points are valid too and hence we can justify the joining of the points (and in this case the pattern suggests a curved line, which we could check by calculating some intermediate points). In practice when the points of a relation graph seem to fall into a distinct pattern (usually a straight or curved line) this is often (but not always, as shown below) an indication that it *would* be appropriate to join points.

A counter-example to this last statement, just to show that care and consideration of individual cases is necessary, is the graph of (number of sides) → (interior angle) of a regular polygon.

Number of sides	3	4	5	6	7	8	9	10	11	12
Interior angle	60°	90°	108°	120°		135°		144°		150°

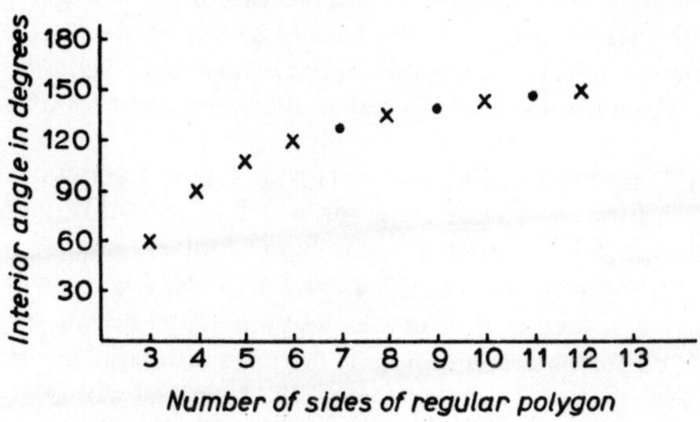

Fig. 14

Although in this case we would be justified in using the pattern to estimate the values of the interior angles of any missing polygons, we would not be correct in joining successive points, as to do so would imply, for example, that we could assign an interior angle to a polygon with $3\frac{1}{2}$ sides!

Occasionally one meets a relation situation where disjointed steps are appropriate. Such an example is the (weight of a letter) → (postage). Children could graph this from up-to-date Post Office charges.

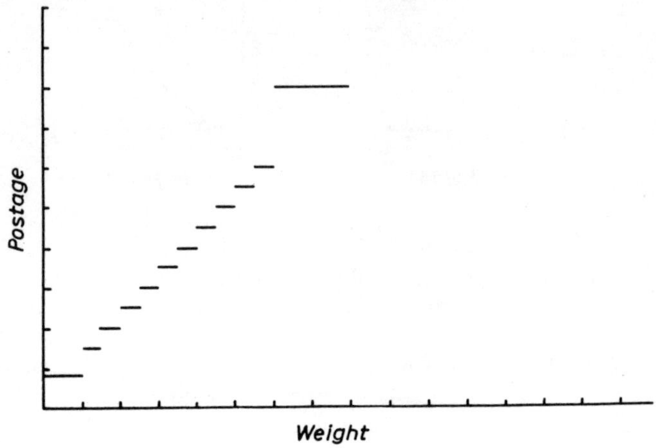

Fig. 15

In a relation graph it is immaterial which of the two sets appears on which axis. In many experimental situations, however, particular members of one set are *chosen* for convenience and the related members of the other set are found. Conventionally the *chosen* set is usually marked on the horizontal axis. For example, in an experiment to investigate the elasticity of a spring by hanging weights on it one might choose particular weights of, say 5 g, 10 g, 15 g, ... and measure the resulting extension. In this case the set of weights would usually be put on the horizontal axis and would be referred to technically as the *independent variable* and the lengths of the extensions would be called the *dependent variable.*

Summary – we have considered two major types of graph drawn on squared paper. First the frequency graph which is a count of members in separate sets and which is made up of columns built up unit by unit as the count proceeds. Second the relation graph which depicts some connection between individual members of one set and individual members of another. Single points on such a graph represent the pairs of related events.

We are by no means suggesting that all the technical words we have used should be learnt by the pupils themselves. Most teachers will probably find *counting graph*, *relation*, *set*, *axis* quite sufficient. We have mentioned other words in common use amongst mathematics teachers, and in the literature, simply to help teachers who are not familiar with them.

Neither do we suggest that it is necessary for pupils to be familiar with all these graphical methods and conventions *before* they become involved in the practical work and presentation of results. Such a mass of theory before any action would be unlikely to create enthusiasm or a co-operative attitude. The teacher should himself be familiar with the basic principles discussed in this chapter so that he can guide his pupils to them using their experiments as the examples.

4

Notes on The Classroom Activities (1)

Activities 1–21

Index to Chapter 4

The activities are not analysed in the order in which they were listed in Chapter 2 as it is more convenient to group them where they have common features. This Index will enable teachers to locate particular experiments.

The first twenty-one activities suggested in Chapter 2 were all *a priori* situations (see p. 7). In other words in all these cases it is possible to arrive at a theoretical measure of probability by analysis of all the possible outcomes.

Many books give the theoretical analysis first and almost as an afterthought (if at all) suggest practical work to follow. We feel that pupils are unlikely to be able to appreciate the theoretical analysis until they have personal experience of the 'behaviour' of any given situation. Hence our approach moves from the practical to the theoretical. It encourages the pupil to explore real events, to look for patterns in the results and to try to find theoretical explanations for those patterns. We believe that *all* those three things are important and hence we have only included experiments where there is a chance that the subsequent theoretical analysis will be comprehensible to some, even if not to all, of the pupils. This has led us to omit some fairly common activities (e.g. Count Buffon's needle) where the theoretical analysis is too sophisticated. Through this approach we hope that some understanding of the difficult relationship between reality and prediction may emerge.

Experiments 1, 7, 14, 21 – coin tossing

1. If you tossed four coins 50 times about how many times do you *think* you would get at least two HEADS in the four tosses? Try it. Then repeat the experiment using just two coins, then three coins. About how often do you think you would get at least two HEADS if you tossed *five* coins 50 times?

7. If you toss five coins many times what number of HEADS out of five do you think is likely to appear most often? What number would you expect least? Do the experiment and record the number of HEADS each time. Can you think of a reason for the pattern of the results?

14.

1					F
				E	
			D		
		C			
	B				
A					2

Two players (1) and (2) take 16 counters each and place them in the squares marked 1 and 2 respectively. Then they take it in turns to toss a coin. On each throw, if (1) tosses a HEAD he moves his first counter one place to the right → and if a TAIL he moves it one place down ↓ . When (2) tosses a HEAD he moves his first counter one place to the left ← and for a TAIL one place up ↑ . This goes on until the first counter reaches any of the squares lettered A, B, C, D, E, F. When a counter gets 'home' then that player starts a new counter. The winner is the player who, at the end, has most counters in squares A and F *together*. If this is a tie then the counters in B and E are counted instead. Play the game several times. Is there a pattern in the results?

21. Toss a coin five times and record the number of HEADS. Toss five more times and record the total number of HEADS out of the ten throws so far. Continue in batches of five throws keeping a 'running total' of the number of HEADS and the number of throws. Try to illustrate the results by a graph.

The actual results obtained by one pupil doing Experiment 1 are shown in Fig. 16 and in the frequency graph (Fig. 17).

Number of coins tossed per trial	4	3	2
Number of trials	50	50	50
Number of trials showing two or more heads	37	25	9

Fig. 16

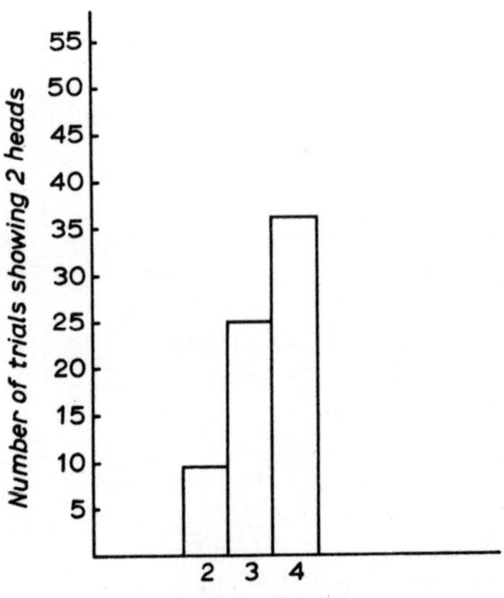

Fig. 17

The pupil predicted, on the basis of these results, that if 5 coins were tossed 50 times then over 40 of the trials would probably show at least 2 HEADS. She then threw 5 coins and obtained the result 44 times. Another pupil only got it 36 times – providing basis for useful discussion. If, in theory, we think about the possible outcomes of tossing two coins we see that we will either get two heads or two tails or one of each. Are these three outcomes equally likely?

A typical result of a two coin experiment is given in the graph below. This graph gives the distinct impression that probably these possibilities are *not* equally likely.

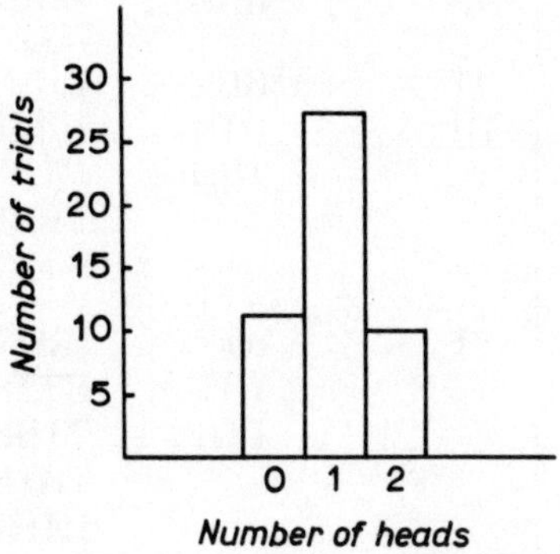

Fig. 18

If we distinguish between the two coins (this is easily done in practice by using two different denominations) then we see that we can classify the results in *four* ways as shown in Fig. 19.

	1st coin	2nd coin
1	H	H
2	H	T
3	T	H
4	T	T

Fig. 19

Experience shows that in the long term the proportion of 2 HEADS, 2 TAILS or 1 of each is as in this table and leads to the *assumption* that it is *these* four outcomes (HH, HT, TH, TT) that are equally likely. This would assert that results with 2 heads or 2 tails are as likely as each other and that results with 1 of each are twice as likely as either of the first two.

Such arguments clearly point to a *numerical* measure of theoretical probability. We say that the theoretical probability of obtaining HH (or HT or TH or TT) in this situation is $\frac{1}{4}$ (or 0·25 or 25%) and so that of

	Number of coins 2	3	4	5
0 TAILS	HH	HHH	HHHH	HHHHH
1 TAIL	HT TH	HHT HTH THH	HHHT HHTH HTHH THHH	HHHHT HHHTH HHTHH HTHHH THHHH
2 TAILS	TT	HTT THT TTH	HHTT HTTH TTHH THTH HTHT THHT	HHHTT HHTTH HTTHH TTHHH HHTHT HTHHT THHHT HTHTH THHTH THTHH
3 TAILS		TTT	TTTH TTHT THTT HTTT	TTTHH TTHHT THHTT HHTTT TTHTH THTTH HTTTH THTHT HTTHT HTHTT
4 TAILS			TTTT	TTTTH TTTHT TTHTT THTTT HTTTT
5 TAILS				TTTTT
Number of outcomes	4	8	16	32

Fig. 20

obtaining a head and a tail is $\frac{1}{2}$ (or 0·5 or 50%). What number would be assigned to absolute certainty? What number would be assigned to absolute impossibility? All other measures of probability lie somewhere between these two numbers. (See Chapter 9.)

We stress the importance of the 'equally likely' hypothesis in this analysis – it underlies all the subsequent experiments in this chapter although we will not deal with it in such detail every time. It was differences of opinion over such assumptions which led to some of the famous arguments about probability in the 17th century. Notice that the hypothesis cannot be proved – we can only adopt the best one to fit the experimental results – another good reason for doing practical work before considering the theoretical situation.

Fig. 20 gives the accepted equally likely outcomes of tossing 2, 3, 4, 5 coins. The dotted lines classify the outcomes according to the number of TAILS. Notice the wealth of pattern in this table.

If the outcomes of each type are counted then an interesting number pattern emerges:

	Number of coins			
	2	3	4	5
0 TAILS	1	1	1	1
1 TAIL	2	3	4	5
2 TAILS	1	3	6	10
3 TAILS		1	4	10
4 TAILS			1	5
5 TAILS				1
Total Number of outcomes	4	8	16	32

Fig. 21

The numbers in this table form part of the so-called 'Pascal's Triangle' which arises in situations involving two-way choices, as this one does in its choice between H and T. This number triangle occurs elsewhere in this book and in *Mathematical Pattern* as well as in many of the references in Appendix 1.

Why does the number of outcomes double as one coin is added to each trial?

Why should the number of '2 tail' outcomes from 5 coins be the sum of the '1 tail' and '2 tail' outcomes of 4 coins?

Is this relationship general throughout the table?

Can you deduce the numbers for 6 coins from the table?

If we reconsider Experiment 1 in the light of this table we see that for 2 coins, 1 out of 4 outcomes contains at least two heads: for 3 coins the figure is 4 out of 8: for 4 coins it is 11 out of 16. Hence we say that the probability of obtaining at least two heads with 2, 3, 4 coins is $\frac{1}{4}$, $\frac{4}{8}$, $\frac{11}{16}$ respectively. These three fractions are easily compared with each other for size on account of their related denominators. How can they best be compared with the *experimental* results of $\frac{9}{50}$, $\frac{25}{50}$, $\frac{35}{50}$?

From such a question we realize the usefulness of decimal fractions and percentages compared to that of vulgar fractions. In this particular case then we have experimental results of 18%, 50% and 74% to compare with theoretical measures of 25%, 50% and 68·75%. It is interesting to note variations in the results obtained by different groups of pupils and then to combine several sets of results as it is likely (but not certain) that the collected results will be closer to the theoretical measure than the individual ones. This idea is of real importance in 'sampling' (see Chapter 8).

In the case of 5 coins in Experiment 1 the theoretical probability of obtaining at least 2 heads would be $\frac{26}{32}$ (use Fig. 21) or 81·25%. The pupil's estimate of over 40 out of 50 trials, based on her own *experimental* results is thus seen to be a reasonable one theoretically.

It seems important that pupils should have the opportunity to make predictions based on their own experiences, to compare their prediction with theoretical measures and then to see how their prediction is matched in further experiment. Thus we may be able to help them to appreciate to some extent how a mathematical model is constructed to explain some facet of the behaviour of the real world. This in turn may give some insight into what some mathematics is about even if their own powers of operation in the field are very limited.

Experiment 7 should be analysed in the light of the previous two tables. The predictions which pupils make *before* doing the experiment may vary widely especially if they have not done any coin-tossing work before. However, we do *not* advise discussion of the possible outcomes before the experiment is done for an ill-informed prediction offers a useful contrast to to better ones made after experience.

In Fig. 22 are the individual results of 25 pupils all tossing 5 coins 32 times (practical note – they did not all do this at the same time, as a whole range of experiments were in progress. Noise in throwing experiments can be drastically reduced by using a plastic pot lined with felt or self-stick baize as a shaker and by throwing into a similarly lined wooden tray or box lid). A total of 32 trials per person was chosen here so that there would be an individual, and immediate, comparison with the $\frac{1}{32}$ $\frac{5}{32}$ $\frac{10}{32}$ $\frac{10}{32}$ $\frac{5}{32}$ $\frac{1}{32}$ probability for 5 coins (see Fig. 21).

Number of trials showing:					
0 Heads	1 Heads	2 Heads	3 Heads	4 Heads	5 Heads
2	6	7	14	3	0
0	10	12	6	4	0
2	5	6	16	2	1
2	2	9	11	6	2
1	3	9	10	7	2
1	3	9	12	5	2
0	2	8	11	11	0
2	3	8	12	6	1
2	6	7	8	7	2
2	8	9	6	6	1
1	6	9	11	5	0
1	7	8	10	5	1
1	4	7	11	7	2
1	5	8	8	4	6
1	5	7	13	6	0
1	4	11	11	3	2
4	6	8	6	7	1
2	5	5	10	10	0
3	8	8	7	3	3
2	3	7	16	4	0
2	5	11	10	4	0
0	4	9	10	8	1
3	7	12	3	4	3
0	5	15	5	5	2
0	6	9	10	5	2
36	128	218	247	137	34
TOTAL 800 trials					
4·5%	16%	27·25%	30·875%	17·125%	4·25%

Fig. 22

We also give the overall experimental probabilities compared with the appropriate theoretical ones (above) expressed as percentages. This is illustrated by the graph (Fig. 24) in which the grey shading represents the experimental results, and the unshaded areas the theoretical ones.

	Number of heads						
	0	1	2	3	4	5	*Total*
Experimental % of trials in each category	4·5	16	27·25	30·875	17·125	4·25	100
Theoretical probability %	3·125	15·625	31·25	31·25	15·625	3·125	100

Fig. 23

(Teachers may prefer to work to the nearest 1% in which case the total may not be exactly 100% – this is worth discussion)

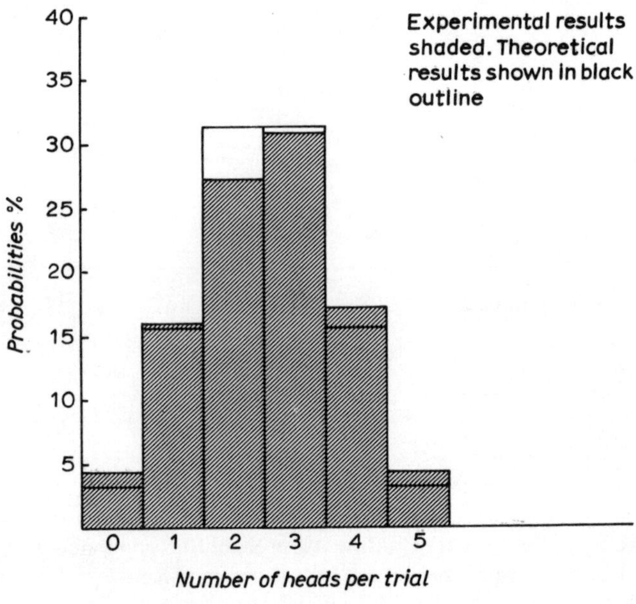

Fig. 24

It is worth discussing the fact that although the individual results probably differ widely from the theoretical probabilities yet the combined results probably do not – if there are enough of them. What is 'enough'? Is it the same number for all experiments? Answers to these questions should begin to emerge as children do more practical work and compare and combine their results.

The general shape of the frequency graph in Fig. 24 should be noted too as it appears often in work on probability and statistics.

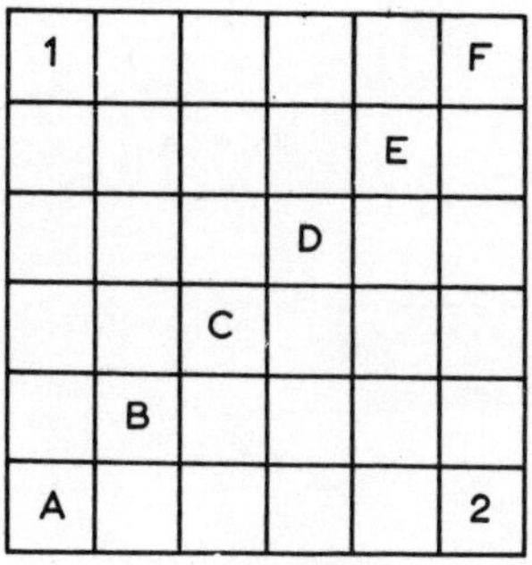

Fig. 25

Experiment 14 is related to *Experiment 7* as each counter takes five moves to get 'home' from its starting-point. Each move is decided by the result of a coin toss. If for example (1) moves a counter to F then the sequence of moves was →→→→→ which *must* have arisen from HHHHH. A counter moving to E, however, could have arisen from any of the five results HHHHT, HHHTH, HHTHH, HTHHH, THHHH.

By similar argument the theoretical probabilities of landing on A, B, C, D, E, F are in the ratio 1 : 5 : 10 : 10 : 5 : 1 – a line from Pascal's Triangle. If the game were played on a (5 x 5) board then the line would be 1, 4, 6, 4, 1. Having played the game several times pupils should be able to predict the most likely 'home' squares and, if they have been involved in other coin-tossing activities already, they may be able to analyse the results in these terms. Alternatively they can be asked to investigate the number of different routes to each of the 'home' squares.

A boy's results for *Experiment 21* are given in Fig. 26.

Running total of heads	2	6	9	11	12	14	16	19	21	23	26	28	31	32	34	37	40	42	45	49
Running total of tosses	5	10	15	20	25	30	35	40	45	50	55	60	65	70	75	80	85	90	95	100
% of heads (nearest %)	40	60	60	55	48	47	46	48	47	46	47	47	48	46	45	46	47	47	47	49

Fig. 26

The graph of the relation (number of tosses) → (% of heads) for a normal coin usually settles quickly to around 50% and provides a nice example of what we mean by 'in the long run'. The graph from Fig. 26 is given below.

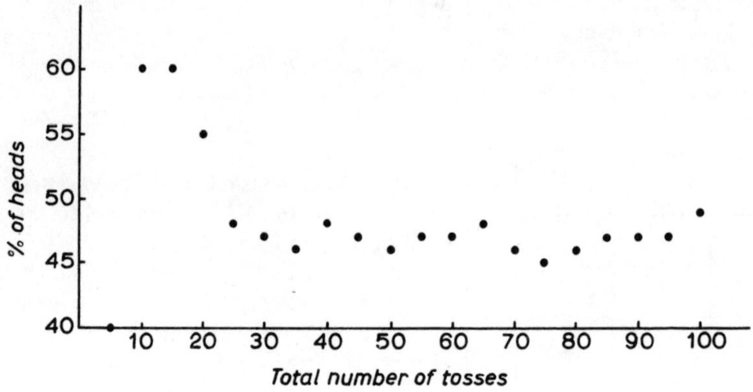

Fig. 27

If there are more trials and the percentage calculated to several decimal places then a graph of (difference between percentage of heads and 50%) → (total number of trials) is often interesting.

Experiments 4, 8, 9, 10, 12, 13, 15

4. If you throw four dice fifty times how often do you think you will get at least one SIX? Try it. Then try it with just one die, then with two dice, then with three. About how often do you think you would throw at least one SIX if you threw five dice fifty times?

8. If you toss two dice many times and add together the two scores each time, which total score do you think you will get most often? What are the highest and lowest total scores possible? Which score do you think will turn up least often? Do the experiment and record all the scores. Can you think of an explanation for the results?

9. If you throw three dice together and add the scores, what are the lowest and highest totals you can possibly get? Which scores would you expect to turn up most often and least often? Do the experiment by throwing three dice at least 300 times. Can you think of an explanation of the results?

10. If you throw three dice together do you think it is most likely that you will get three SIXES or two SIXES or one SIX or no SIXES? Throw the three dice about 300 times and record the number of sixes each time.

12. If you throw two dice many times and record the *difference* between the two scores, what is the smallest possible difference? What is the largest possible difference? Which differences do you think will turn up most often and least often? Throw the two dice about 100 times. Are the results as you expected? Can you explain them?

13. One of these two dice is biased (that means it is an unfair die). Can you find out, by some tossing experiments, which one it is and to which number or numbers it is biased?

15. If you toss two dice many times and *multiply* the two scores together do you think that it is equally likely that the answer you get is exactly divisible by three, or gives remainder 1 when divided by three, or gives remainder 2 when divided by three? Throw the dice at least 100 times and after multiplying the two numbers, divide by three and record the *remainder.* Was your prediction correct? Can you find an explanation for the results?

These experiments all use cubical dice. It is assumed that if the die is fair then each of the six numbers has an equal chance of appearing on top. The experiments suggested can be varied considerably. A die need not necessarily be a cube but can be made in the shape of any other convex regular solid (tetrahedron, numbered 1 to 4; octahedron, numbered 1 to 8; dodecahedron, numbered 1 to 12; and icosahedron, numbered 1 to 20). *Regular* solids are necessary to preserve the equal likelihood assumption. Another alternative is a card or perspex regular polygon with a central hole and pointed shaft to make a 'spinner'. In an experiment involving several dice it may be easier for some pupils to work out all the possible outcomes if the dice are distinguishable by colour and/or size.

If one die is thrown then one set of possible equally likely outcomes is {1, 2, 3, 4, 5, 6}. If the two dice are thrown then we can list a set of 36 (6 x 6) 'ordered pairs' thus:

{(1, 1)	(2, 1)	(3, 1)	(4, 1)	(5, 1)	(6, 1)
(1, 2)	(2, 2)	(3, 2)	(4, 2)	(5, 2)	(6, 2)
(1, 3)	(2, 3)	(3, 3)	(4, 3)	(5, 3)	(6, 3)
(1, 4)	(2, 4)	(3, 4)	(4, 4)	(5, 4)	(6, 4)
(1, 5)	(2, 5)	(3, 5)	(4, 5)	(5, 5)	(6, 5)
(1, 6)	(2, 6)	(3, 6)	(4, 6)	(5, 6)	(6, 6) }

Notice that (1, 6) and (6, 1) represent different equally likely outcomes in this set as the numbers appear on different dice (this is easily realized if the dice are different colours) hence the term *ordered* pairs.

We could tabulate 216, (6 x 6 x 6), outcomes of throwing three dice as ordered triples (1, 1, 1), (1, 1, 2), etc., but the tabulation for this and even more dice would be very tedious. In this and many other situations a 'tree diagram' is a useful device for showing up the structure of outcomes. Fig. 28 shows such a diagram for outcomes of dice throwing.

Pupils working on *Experiments 4 and 10* could make a large scale chart for up to 4 dice, marking the branches showing a SIX in colour so that they are easily picked out.

Fig. 29 is built up from just such a diagram and shows the outcomes classified according to the number of SIXES they contain. Compare this with the Fig. 21 for coin tossing on page 31.

What pattern can you see here? Notice that all the numbers in the table are multiples of powers of 5 (i.e. $5^1, 5^2, 5^3, 5^4 \ldots$), even the number 1 which is 5^0 (see *Number Appreciation*). The table in Fig. 30 is the one in Fig. 29 with all the numbers written in that way.

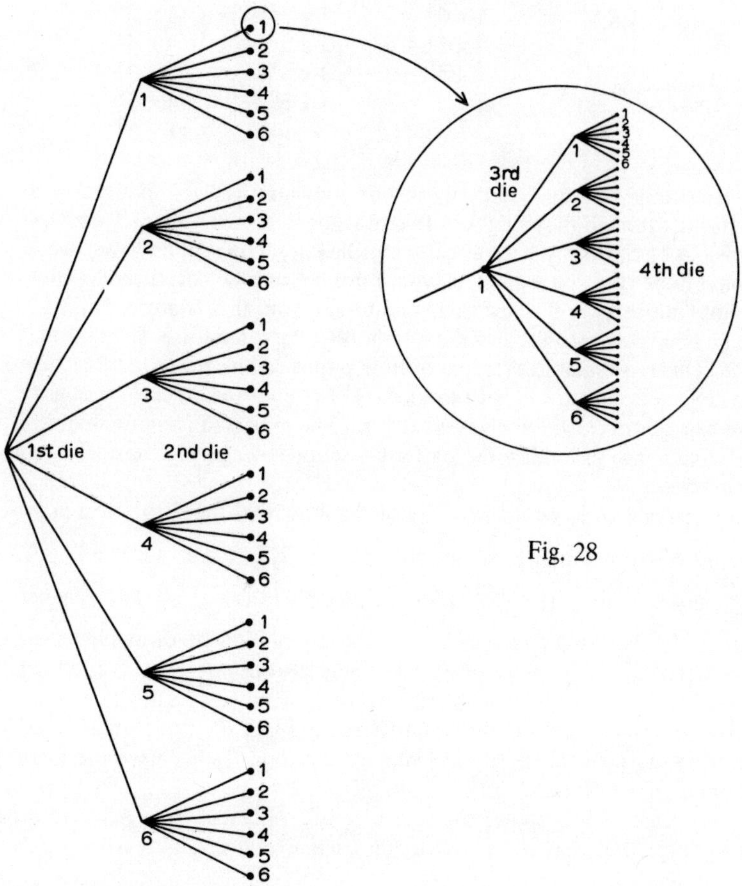

Fig. 28

	Number of dice				
	1	2	3	4	5
0 sixes	5	25	125	625	
1 sixes	1	10	75	500	
2 sixes		1	15	150	
3 sixes			1	20	
4 sixes				1	
5 sixes					
Total	6	36	216	1,296	

Fig. 29

1×5^1	1×5^2	1×5^3	1×5^4
1×5^0	2×5^1	3×5^2	4×5^3
	1×5^0	3×5^1	6×5^2
		1×5^0	4×5^1
			1×5^0

Fig. 30

We should be surprised if any of our pupils noticed the 'powers of 5' pattern, although they might notice that the numbers, except 1, are divisible by 5. If they have already met Pascal's Triangle, however, we would hope that some of them would notice that the coefficients (multiplying numbers) in this last table are in fact from that triangle. This should make it very easy to predict the numbers in the table for 5, 6, or more, dice. This is a valuable example of how pupils can be led to see that the existence of pattern can help to unravel new situations. *Teachers* might like to explore for themselves what would be expected from non-cubical dice and hence generalize the pattern – it appears again in the analysis of *Experiment 19.*

In relation to *Experiment 4* the tables show that the theoretical probabilities of throwing at least one SIX with 1, 2, 3, 4 dice are $\frac{1}{6}$, $\frac{11}{36}$, $\frac{91}{216}$, $\frac{671}{1296}$ respectively, i.e. 17%, 31%, 42%, 52% to the nearest 1%. Teachers can verify that in the case of 5 dice the probability of throwing at least one SIX is approximately 60%. They may also discover, by continuing the tables, that in this case the probability of no sixes and the probability of exactly one six are equal. If the number of dice is increased to six then the probability of throwing exactly one six becomes the highest, as one might expect.

The graph below shows the actual results obtained by a girl who used them to predict that about 32 of the trials would give at least one SIX

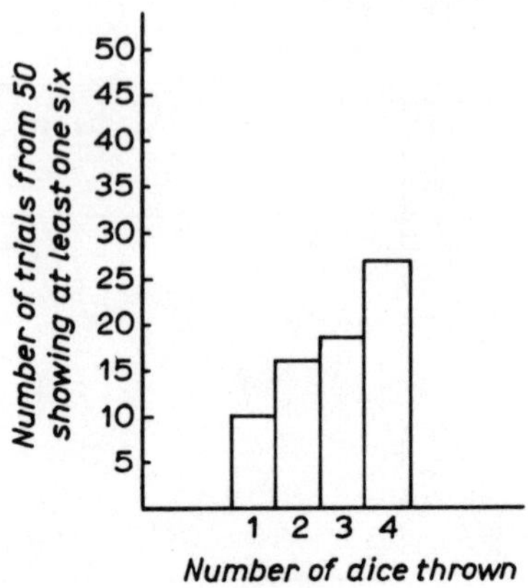

Fig. 31

if she threw five dice 50 times. Compare this result and prediction with the theoretical probabilities given above.

Experiment 10 was organized by a group on a large scale. The theoretical probabilities of throwing 0 sixes, 1 six, 2 sixes, 3 sixes with 3 dice are $\frac{125}{216}$, $\frac{75}{216}$, $\frac{15}{216}$, $\frac{1}{216}$ respectively (taking the appropriate line from Fig. 29).

Experimental results and comparison with theoretical probability are given in Fig. 32 and again it was noticed how close these 'long run' results were compared with the individual experiments from which they were obtained.

	Number of sixes thrown using three dice			
	0	1	2	3
Number of trials . . .	3,202	1,810	410	50
Percentage of total trials	58·5%	33·1%	7·5%	0·9%
Theoretical probability .	57·9%	34·7%	6·9%	0·5%

Fig. 32

Experiments 8 and 9 concern the total scores from 2 or 3 dice. It is unusual for pupils (or adults for that matter) to make good predictions about the relative occurrence of the various scores before the experiment although many will say that they think that the minimum and maximum scores (2, 12 and 3, 18 respectively) are rare.

A pupil, within our range, who thought that all the scores except perhaps 2 and 12 would come up "about even" threw two dice 150 times with the following result.

Score	2	3	4	5	6	7	8	9	10	11	12
Number of trials	2	12	13	15	21	26	19	19	10	8	5

Fig. 33

He concluded that the scores got progressively more likely up to 7 and then the probabilities went down again. This pupil was able to follow this up with an explanation based on the possible outcomes by listing them all (see page 38).

If we construct a two-way entry table, or matrix, showing the ways in which the various scores occur we get:

	+	Score on 2nd die 1	2	3	4	5	6
Score on 1st die	1	2	3	4	5	6	7
	2	3	4	5	6	7	8
	3	4	5	6	7	8	9
	4	5	6	7	8	9	10
	5	6	7	8	9	10	11
	6	7	8	9	10	11	12

The numbers within the rectangle are the totals.

Fig. 34

We can deduce that the theoretical probabilities of obtaining each score from 2 to 12 are proportional to the number of times that score appears in the table, giving:

Total score . .	2	3	4	5	6	7	8	9	10	11	12
Probability . .	$\frac{1}{36}$	$\frac{2}{36}$	$\frac{3}{36}$	$\frac{4}{36}$	$\frac{5}{36}$	$\frac{6}{36}$	$\frac{5}{36}$	$\frac{4}{36}$	$\frac{3}{36}$	$\frac{2}{36}$	$\frac{1}{36}$

Fig. 35

What is it about the dice that makes the probabilities start to decrease after the score of 7?

The possible scores for 3 dice are easily obtained from the matrix for 2 dice by adding the six possible scores of the 3rd die as shown opposite.

	3	4	5	6	7	8		4	5	6	7	8	9
	4	5	6	7	8	9		5	6	7	8	9	10
add 1	5	6	7	8	9	10	add 2	6	7	8	9	10	11
	6	7	8	9	10	11		7	8	9	10	11	12
	7	8	9	10	11	12		8	9	10	11	12	13
	8	9	10	11	12	13		9	10	11	12	13	14
	5	6	7	8	9	10		6	7	8	9	10	11
	6	7	8	9	10	11		7	8	9	10	11	12
add 3	7	8	9	10	11	12	add 4	8	9	10	11	12	13
	8	9	10	11	12	13		9	10	11	12	13	14
	9	10	11	12	13	14		10	11	12	13	14	15
	10	11	12	13	14	15		11	12	13	14	15	16
	7	8	9	10	11	12		8	9	10	11	12	13
	8	9	10	11	12	13		9	10	11	12	13	14
add 5	9	10	11	12	13	14	add 6	10	11	12	13	14	15
	10	11	12	13	14	15		11	12	13	14	15	16
	11	12	13	14	15	16		12	13	14	15	16	17
	12	13	14	15	16	17		13	14	15	16	17	18

Fig. 36

These totals may be analysed to give probabilities as follows:

Total score	3	4	5	6	7	8	9	10	11	12	13	14	15	16	17	18
Probability	$\frac{1}{216}$	$\frac{3}{216}$	$\frac{6}{216}$	$\frac{10}{216}$	$\frac{15}{216}$	$\frac{21}{216}$	$\frac{25}{216}$	$\frac{27}{216}$	$\frac{27}{216}$	$\frac{25}{216}$	$\frac{21}{216}$	$\frac{15}{216}$	$\frac{10}{216}$	$\frac{6}{216}$	$\frac{3}{216}$	$\frac{1}{216}$

Fig. 37

The table (Fig. 38) shows successively the results (as a percentage of the total number of throws) obtained by one pupil, by a class of twenty, and by a combined effort involving eight classes. All figures are to the nearest 0·5%. Notice the changing differences between the experimental and theoretical results as the number of trials increases.

Score with 3 dice	1 pupil	1 class	8 classes	Theoretical probability
	%	%	%	%
3	0	0	0·5	0·5
4	0	1·5	1·5	1·5
5	3·0	1·5	2·5	3·0
6	0	4·0	4·5	4·5
7	11·0	6·0	7·5	7·0
8	14·0	9·0	9·5	9·5
9	22·0	12·5	11·5	11·5
10	14·0	14·5	12·5	12·5
11	8·5	12·5	12·5	12·5
12	5·5	11·5	11·5	11·5
13	8·5	10·0	9·0	9·5
14	0	8·0	7·0	7·0
15	14·0	5·0	5·0	4·5
16	0	2·0	2·5	3·0
17	0	1·5	1·0	1·5
18	0	0·5	1·0	0·5

Fig. 38

These results are given as graphs to illustrate more vividly the comparison of experimental results and theoretical probabilities (see Figs. 39 and 40 on pages 46 and 47).

Experiments 12 and 15 concern two dice and the difference and product of the individual scores. (The *product* of two numbers is the result of multiplying one by the other.)

The first two tables (Figs. 41 and 42) show the possible differences and their respective probabilities.

Why does each score appear an *even* number of times? Notice how a graph of the difference probabilities would be non-symmetrical, skewed to one side (Fig. 43).

		2nd die					
		1	2	3	4	5	6
	1	0	1	2	3	4	5
1st	2	1	0	1	2	3	4
die	3	2	1	0	1	2	3
	4	3	2	1	0	1	2
	5	4	3	2	1	0	1
	6	5	4	3	2	1	0

Fig. 41

Hence the differences and their theoretical probabilities are

Difference	0	1	2	3	4	5
Probability	$\frac{6}{36}$	$\frac{10}{36}$	$\frac{8}{36}$	$\frac{6}{36}$	$\frac{4}{36}$	$\frac{2}{36}$

Fig. 42

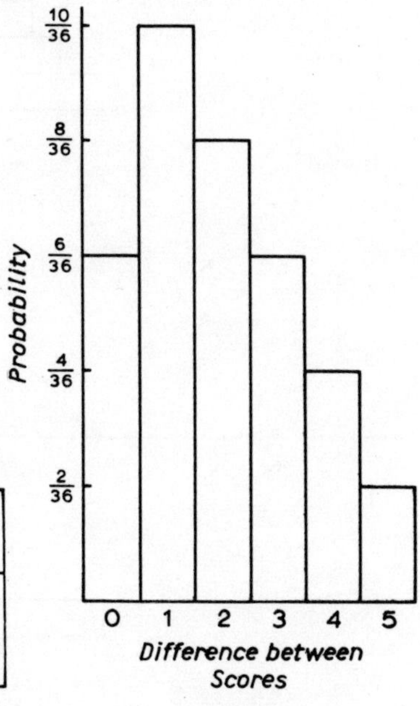

Fig. 43

The next two tables show, respectively, the products of the possible throws and the results of dividing those products by 3.

Products of scores		2nd die 1	2	3	4	5	6
	1	1	2	3	4	5	6
1st	2	2	4	6	8	10	12
die	3	3	6	9	12	15	18
	4	4	8	12	16	20	24
	5	5	10	15	20	25	30
	6	6	12	18	24	30	36

Remainders on division by 3		2nd die 1	2	3	4	5	6
	1	1	2	0	1	2	0
1st	2	2	1	0	2	1	0
die	3	0	0	0	0	0	0
	4	1	2	0	1	2	0
	5	2	1	0	2	1	0
	6	0	0	0	0	0	0

Fig. 44

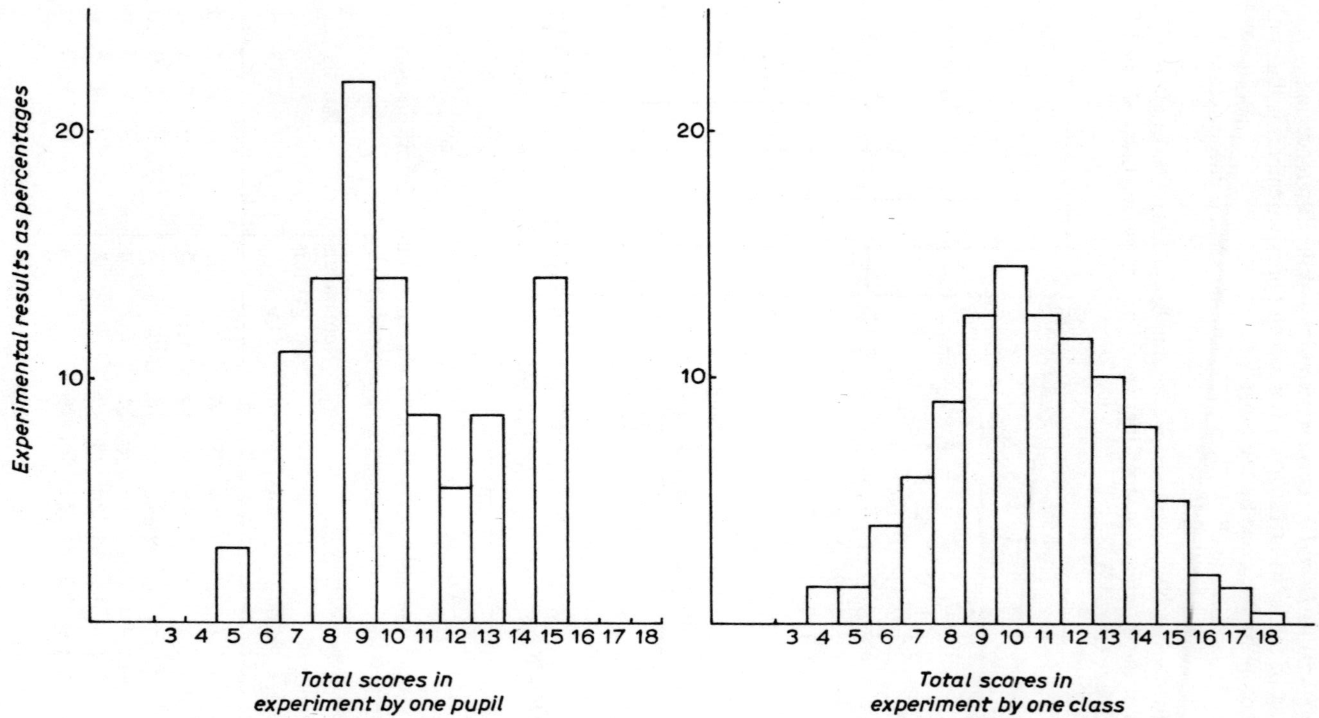
Experimental results as percentages
20
10
3 4 5 6 7 8 9 10 11 12 13 14 15 16 17 18
Total scores in
experiment by one pupil
20
10
3 4 5 6 7 8 9 10 11 12 13 14 15 16 17 18
Total scores in
experiment by one class

(a) (b)

Fig. 39

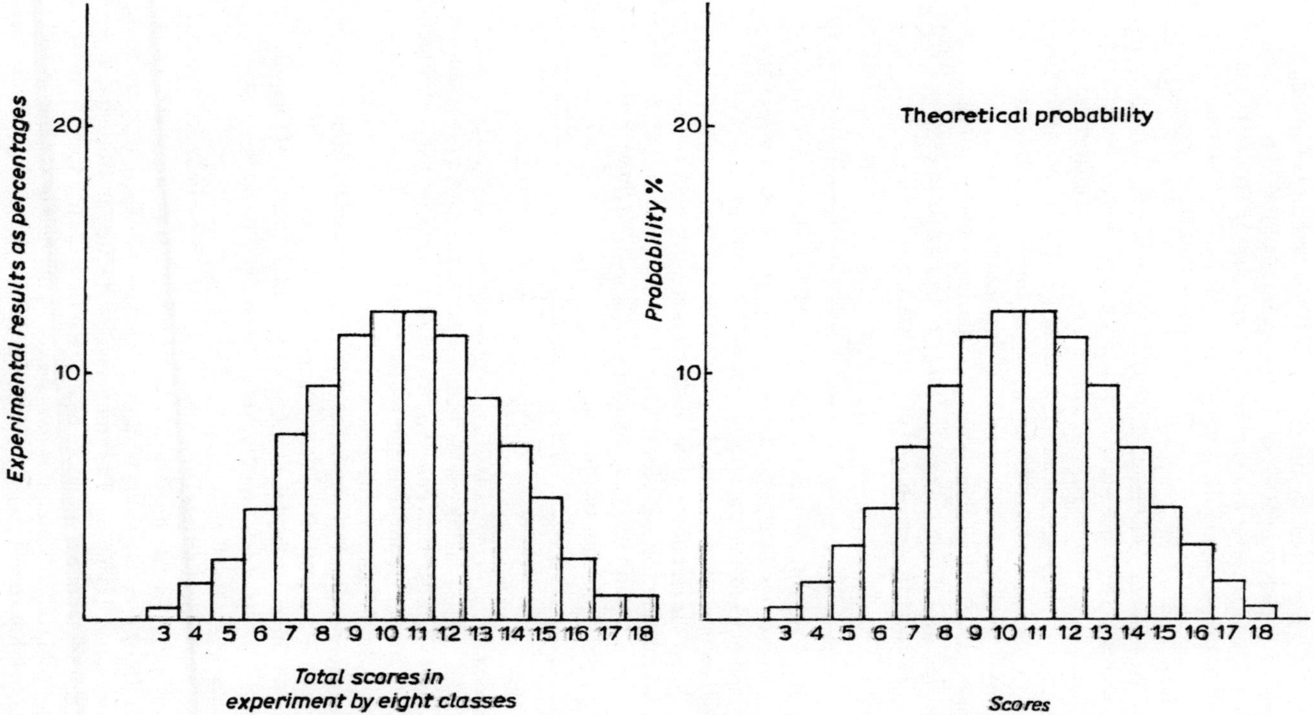
Experimental results as percentages
20
10
3 4 5 6 7 8 9 10 11 12 13 14 15 16 17 18
Total scores in
experiment by eight classes
Theoretical probability
Probability %
20
10
3 4 5 6 7 8 9 10 11 12 13 14 15 16 17 18
Scores

(c) (d)

Fig. 40

It is easily seen from the last table that the probability of getting a multiple of three is over 50% (as '0' appears in 20 of the 36 positions) which may be a surprise to some. Is this result only applicable to the numbers 1 to 6 or would the probability be the same for any pair of random numbers?

For *Experiment 13* a biased die is used. A variety of these may be made in the workshop by embedding metal in a wooden cube. If this is not possible then a hollow cardboard die with several layers of heavy card stuck inside one face produces a reasonable substitute. This activity usually sparks off discussion about the fairness or unfairness of a die and at what stage one is being reasonable in making assumptions about it. Such discussions can only be qualitative not quantitative as the statistical techniques involved are beyond the scope of this work. Interest is added if a number of differently biased dice are used.

11. There are 100 cubes in the box and each has one marked face. If you throw them onto the table how many do you think will land marked side uppermost? Do it and record the number. Then remove the cubes which have fallen marked side up and throw the *rest* down again. Keep on doing this until you have no cubes left. Repeat the experiment from the start several times. About how many cubes do you *expect* to show their marked side after the third throw?

Experiment 11 requires 100 small cubes with one marked face. Small dice could be used taking one particular score as the marked face, but this would be expensive. Unit cubes from a set of Dienes Multibase Arithmetic blocks or from a set of Cuisenaire rods are suitable if one face is marked with felt-pen, in fact Cuisenaire produce such cubes already marked. Alternatively, a group of pupils can make a set themselves from accurately cut square section wood.

Fig. 45 shows firstly the results of a single experiment and then the combined results of ten experiments using the Arithmetic Mean (see Averages, Chapter 10).

This experiment differs from others discussed so far in a very important respect for in this case the results of successive trials *are dependent upon the results of previous trials.* With the coin and dice experiments each trial has been a new beginning. The point is worth discussion with pupils – suppose a fair coin has produced heads each time for five tosses – most people feel intuitively that it is more likely to fall tails next time. In fact the probability of this event remains 50% despite the previous results, but

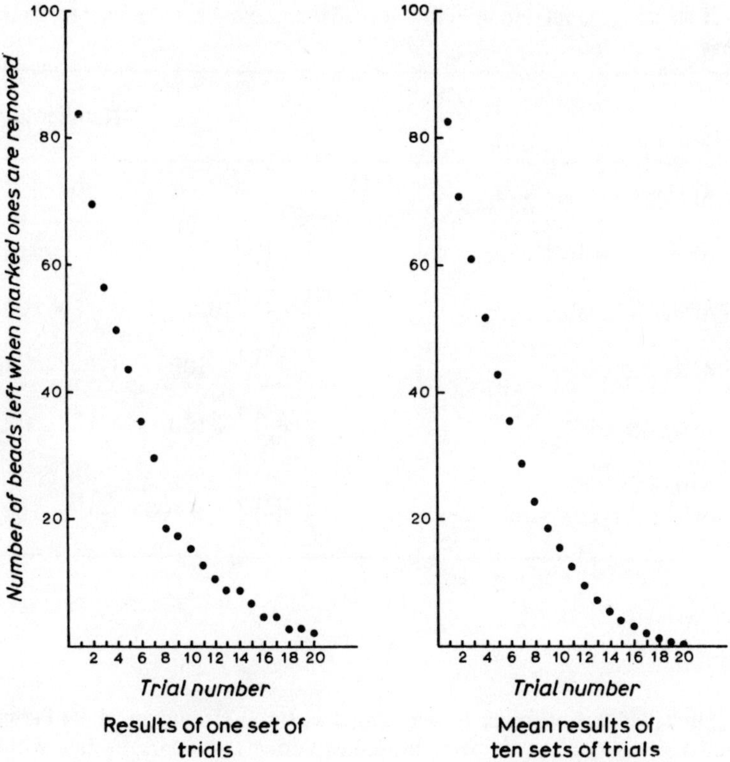

Fig. 45

a desire to 'redress the balance' makes some people assume that the coin can in some way 'remember' what it did before. In this cube throwing experiment, however, the set of cubes thrown is, in a sense, made to 'remember' by the removal of a certain sub-set.

If we assume that the probability of any one cube showing its marked face is $\frac{1}{6}$ then we can assume that, in theory, one sixth of the total number cubes thrown will be removed after any one trial. In theory then we expect to remove $\frac{100}{6}$ cubes after the first throw (but 100 is not exactly divisible by 6 . . . . . .) and have $\frac{5}{6}$ of 100 cubes left to throw again. After the second throw therefore, we expect to remove $\frac{1}{6}$ of ($\frac{5}{6}$ of 100) cubes and have $\frac{5}{6}$ of ($\frac{5}{6}$ of 100) left.

After the 3rd throw we would expect to remove –

$\frac{1}{6}$ of [$\frac{5}{6}$ of ($\frac{5}{6}$ of 100)] cubes and have $\frac{5}{6}$ of [$\frac{5}{6}$ of ($\frac{5}{6}$ of 100)] left.

If all this is tabulated as below then the emerging pattern becomes very clear.

	Number to remove	Number left
At start		100
After 1st trial	$\frac{1}{6} \times 100$	$\frac{5}{6} \times 100$
After 2nd trial	$\frac{1}{6} \times \frac{5}{6} \times 100$	$\left(\frac{5}{6}\right)^2 \times 100$
After 3rd trial	$\frac{1}{6} \times \left(\frac{5}{6}\right)^2 \times 100$	$\left(\frac{5}{6}\right)^3 \times 100$
After 4th trial	$\frac{1}{6} \times \left(\frac{5}{6}\right)^3 \times 100$	$\left(\frac{5}{6}\right)^4 \times 100$
After *n*th trial (where *n* is any whole number)	$\frac{1}{6} \times \left(\frac{5}{6}\right)^{n-1} \times 100$	$\left(\frac{5}{6}\right)^n \times 100$

Fig. 46

Similar experiments can be performed with coins (removing those falling heads), drawing pins (removing those head down), spinners, or dice in the form of regular solids other than cubes, etc.

A few pupils may be able to appreciate that if the experiment is performed with T objects (instead of with 100) with a probability p of the marked side coming up (instead of $\frac{1}{6}$), then, in the expressions given above $\frac{1}{6}$ would be replaced by p and $\frac{5}{6}$ by $(1-p)$ and 100 by T so that in theory the state after n trials would be

$$p \times (1-p)^{n-1} \times T \text{ removed and } (1-p)^n \times T \text{ left.}$$

Of course we do not suggest that pupils be asked to derive or to learn the derivation of such a formula. We mention it because some pupils may be helped to appreciate the power of a general formula by seeing it develop for themselves.

The pattern obtained in these experiments is related to many forms of decay and growth, in particular to radio-active decay and to growth at compound interest. (See some of the books in Appendix 1.) Both these topics could be explored further by selected pupils.

5.	If you have five playing cards including just one ACE and you choose one card, and then replace it, fifty times, how many times do you think you will choose the ACE? Try it. Then repeat the experiment but draw two cards at a time, then draw three cards at a time. About how often do you think you would get the ACE if you drew four cards at a time? How often if you drew five cards?

Experiment 5, using five playing cards one of which is an Ace, is one where a pattern usually emerges very clearly. The graph below shows a set of experimental results compared with theoretical probabilities.

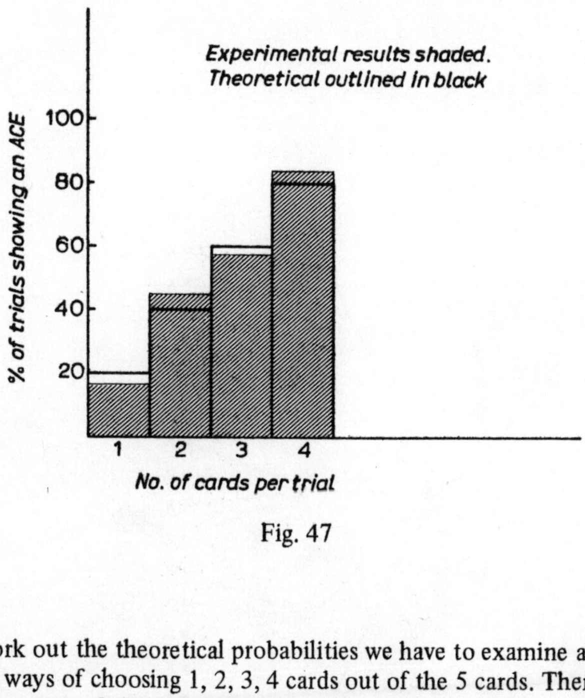

Fig. 47

To work out the theoretical probabilities we have to examine all the different ways of choosing 1, 2, 3, 4 cards out of the 5 cards. There are 5 different ways of choosing 1 from 5 cards and, as the choice of 1 card leaves 4 cards behind, there must also be 5 ways of choosing 4 cards from 5. Similarly the number of ways of choosing 2 cards from 5 is the same as the number of ways of choosing 3 from 5.

A 5-point diagram, Fig. 48, shows the possible ways of choosing 2 cards from 5 cards very clearly. It shows that there are 10 possible pairs.

Each vertex represents a card and each line represents a pairing. This type of diagram is almost always useful in a pairing situation. If we choose one particular vertex as the Ace we see that 4 of the 10 pairings contain the Ace and 6 do not. Therefore, the theoretical probability of getting the Ace is 40% when 2 cards are chosen from 5. Alternatively a tree diagram can be used. This appears to give 20 outcomes, but each pairing appears twice here so there are 10 pairs. For this reason the tree diagram is less useful than the point diagram in this situation.

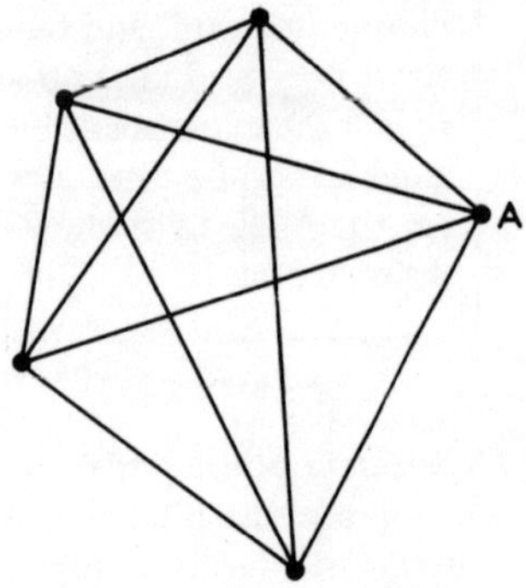

Fig. 48

The choice of 3 cards at a time can be seen clearly on the five point diagram as the ten *triangles* whose vertices are three of the five points.

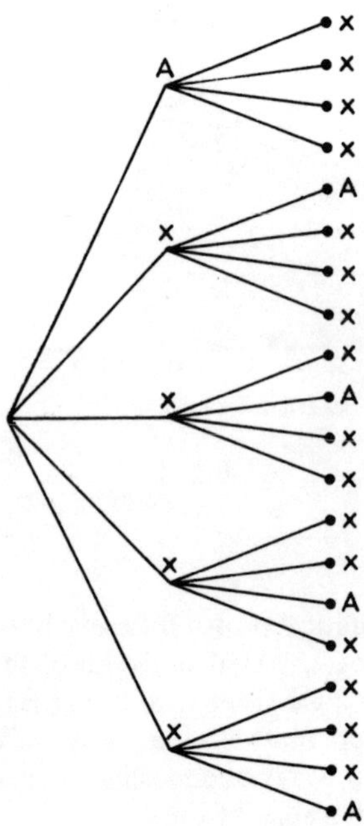

Fig. 49

So many problems in probability and other branches of mathematics involve the number of ways of choosing a sub-set from a set that a symbolic shorthand is used for brevity. We describe it not for the use of pupils but for the information of teachers who are likely to meet it in their own reading on the topic.

The number of ways of choosing 2 things from 5 things is called the 'number of combinations of 2 from 5' and is symbolised as ${}^5C_2$, or ${}_5C_2$ by some writers.

We have already seen from Experiment 5 that ${}^5C_2 = {}^5C_3$ and that ${}^5C_1 = {}^5C_4$. In general, for any pair of numbers n and r, where n is at least as great as r (in mathematical symbols we write $n \geqq r$) then the statement ${}^nC_r = {}^nC_{n-r}$ is true. In Chapter 6 we consider the way of computing ${}^nC_r$ in all cases.

Further study of the coin-tossing table Fig. 20 on page 30 reveals that the Pascal's Triangle numbers in the table Fig. 21 on page 31 could be rewritten:

${}^2C_0$	${}^3C_0$	${}^4C_0$	${}^5C_0$
${}^2C_1$	${}^3C_1$	${}^4C_1$	${}^5C_1$
${}^2C_2$	${}^3C_2$	${}^4C_2$	${}^5C_2$
	${}^3C_3$	${}^4C_3$	${}^5C_3$
		${}^4C_4$	${}^5C_4$
			${}^5C_5$

Using the known relationships between the numbers of Pascal's Triangle we can find many general relationships between combinations. We leave these as investigations for teachers!

Experiments 16, 17, 18, 19

16. Take a box containing one bead of one colour, three of another colour and five of a third colour. If without looking you choose one bead at a time and then replace it when you have seen its colour so that there are always nine in the box, which colour do you think you are likely to choose least often? Which do you think will come out most often? How do you think the two numbers will compare? Now do the experiment many times. Were you about right?

17. Take a box with equal numbers of red and blue beads. Choose one bead at random, record the colour and put it back. Then repeat for a second bead. Do you think you are more likely to draw the same colour bead each time or a different colour? Choose two beads at least 100 times and record the results in columns headed RR, BB, RB. Can you explain the results?

18. Take a box with equal numbers of red and blue beads. Choose one bead at random, record the colour and replace. Do this at least 50 times. After this choose one, replace and choose another and record the colour sequence RR, RB, BR, or BB – repeat this at least 50 times. Lastly do the experiment with three beads, replacing each one directly you have noted its colour so that there are always the same number of beads left in the box.

19. Take a box of equal numbers of red, blue and yellow beads. Choose one at random many times, record the colour and replace. Next choose one, replace and choose another and record the colour sequence RR, RY . . . . . etc. Do this many times. After that repeat the experiment for a run of three beads replacing each bead before choosing another so that there are always the same number of beads to choose from. Can you see and explain any pattern in the results?

Although beads are suggested for these experiments and are easily available it is perfectly possible to substitute pegs, counters, wooden sticks, etc. provided that the objects are identical in size, shape and material so as to eliminate the bias which variations in those properties would produce. For use with beads *sampling bottles* can easily be contrived from a bottle

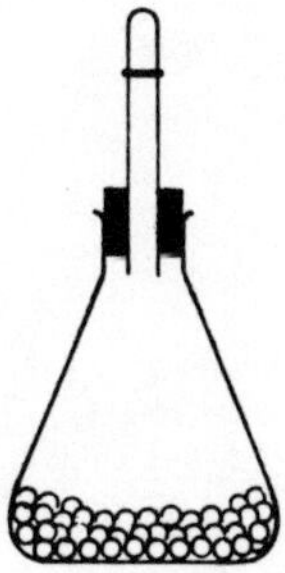

Fig. 50

or flask, cork and hollow tube closed at one end and using an elastic band as an adjustable marker on the glass tube to indicate the sample size (in cases where more than one bead can be chosen at the same time).

Experience of the behaviour of samples taken from a known population is a good basis for understanding the techniques of sampling used in advertising, industry, science and government, and for this reason we include such activities in our suggestions.

Experiment 16 is likely to produce good predictions from pupils and results which compare well with theoretical probability. To take just one example, one pupil sampled one bead 100 times from a bottle containing 1 yellow, 3 blue and 5 red beads. The results were yellow drawn 10 times, blue 32 times and red 58 times. The graphs in Fig. 51 show the pupil's results and the theoretical probabilities of each event (to the nearest 1%) assuming that each bead (not each colour) has an equal chance of being chosen.

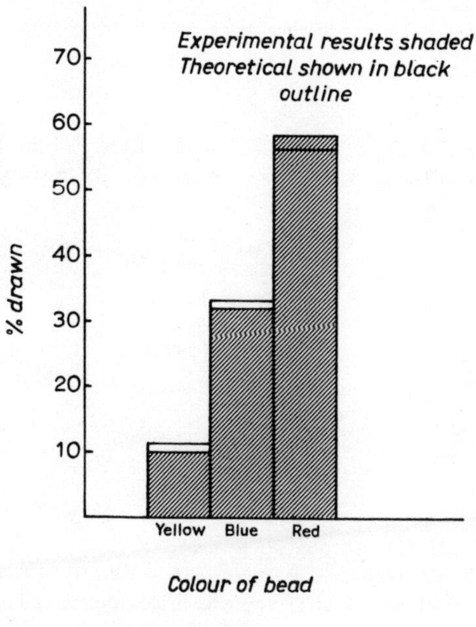

(a)

Fig. 51

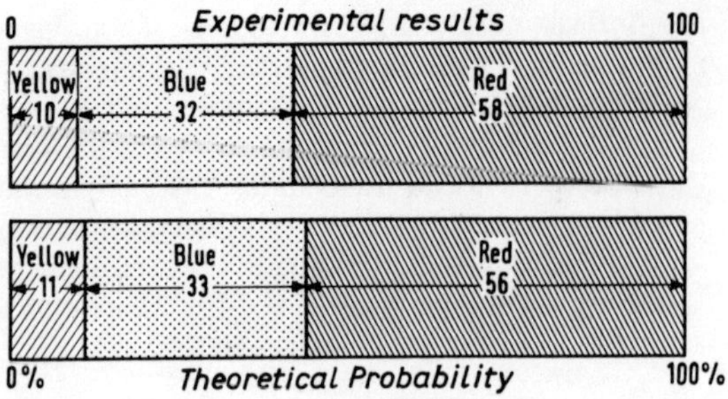

(b)

Fig. 51

Some actual results of *Experiment 17* are as follows:

Number of runs of 2 beads taken	120
Number showing both red	32
Number showing both blue	29
Number showing one of each	59

Fig. 52

In theory one would expect, from 120 trials, 30 trials both red, 30 both blue and 60 one of each. A tree diagram analyses the possibilities:

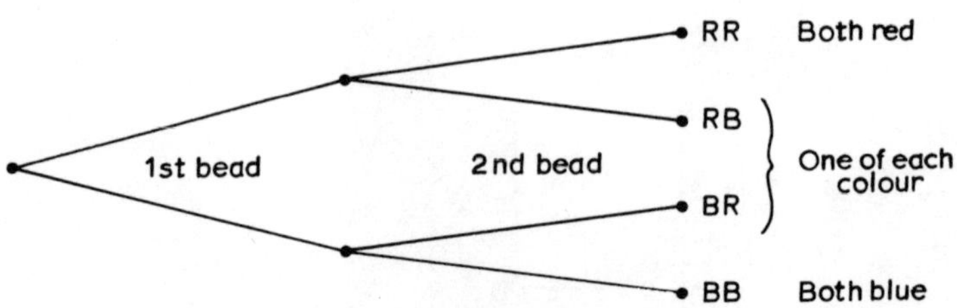

Fig. 53

It is clear that this situation is analogous to that of tossing two coins and recording HEADS or TAILS and one hopes that pupils will notice this. Such observations are worth encouraging as they help pupils to see that one 'mathematical model' can often illustrate a number of seemingly different situations.

Similarly *Experiment 18* produces results analogous to tossing 1, 2, 3, 4 coins and analysis of the possible outcomes gives rise to Pascal's Triangle again (see page 31).

Experiment 19 usually demands well over 100 runs each time before the pattern of results becomes obvious and it is best therefore to have several children or groups working at it. A tree diagram for all the possibilities is similar to that for dice throwing (and coin tossing) and has three branches arising at each junction.

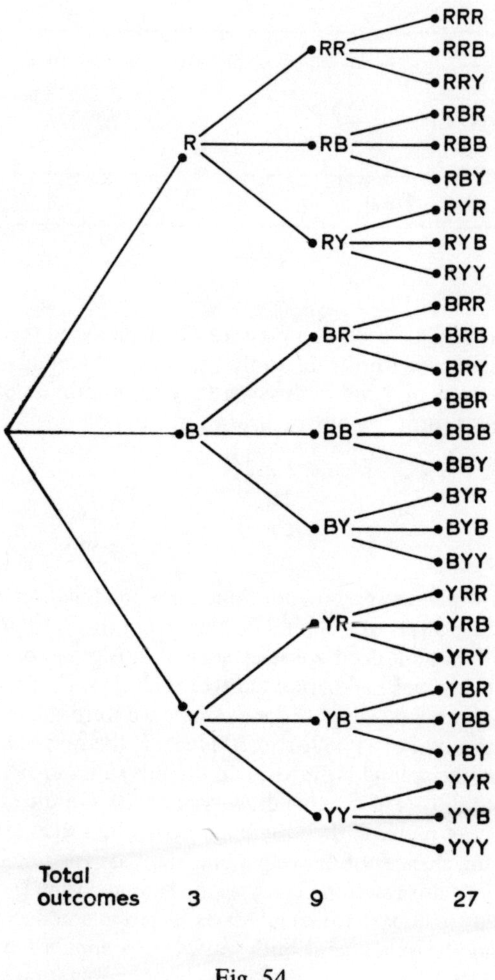

Fig. 54

Thus with one bead the theoretical probabilities of each colour are equal. With runs of two beads the probability of both being the same colour is $\frac{3}{9}$, for three beads the probability of all one colour is $\frac{3}{27}$. Probabilities of two one colour and one another etc. may be found quite easily from the tree diagram.

If we classify the possible outcomes by the number of reds occurring, for example, then the following table emerges:

	Number of beads in sample		
	1	2	3
0 reds	2	4	8
1 red	1	4	12
2 reds		1	6
3 reds			1
Total	3	9	27

Fig. 55

This table is similar to that on page 39 which analysed the number of sixes obtainable in dice throwing. In that situation the numbers were all multiples of powers of 5 and in this case they are multiplies of powers of 2. Written in that form a familiar pattern turns up:

1×2^1	1×2^2	1×2^3
1×2^0	2×2^1	3×2^2
	1×2^0	3×2^1
		1×2^0

The way in which the pattern continues as a combination of Pascal's Triangle and the powers of 2 should be clear from this tabulation.

Teachers will have noticed our insistence, in this group of experiments, on each bead being replaced before another is chosen. Useful discussion can arise over this point. Suppose for example we were to take runs of two beads from a box of 10 reds and 10 blues. If the first one is red and is replaced before the second is drawn then there is still a so-called '50-50 chance' of a red or of a blue being drawn next time. On the other hand, if the red bead is not replaced then there are more blues than reds left and consequently the chances of drawing a blue next are fractionally higher than 50%. Clearly this fraction decreases as the population (i.e. the total number of beads to choose from) increases. A group could explore this further by doing the experiment both ways with a population of 2, 4, 6, . . . beads to try to decide for which minimum population size they feel they can ignore the 'replacement rule'. In theory of course it can never be neglected but in practice there are times when accuracy is sacrificed for speed – another opportunity for useful discussion, not necessarily about mathematics.

If a sampling bottle is used in such an experiment the pupils will have an operational model of the difference between the replacement and non-replacement situations. In the former only the first bead in the tube is recorded and a run of, say, three beads needs three operations of the bottle. For the non-replacement situation, however, the first three beads from one operation are recorded.

2. Do you think that it is likely that two people in the class have the same birthday? Find out. Get information about this from other classes as well.

3. Do you think that out of five people two at least are likely to have the same birth month? Write down the birth months of fifty sets of five people. In how many of your sets were at least two people born in the same month?

20. Make a row of the twelve boxes which each contain all the letters of the alphabet. Starting from one end of the row take one letter at random from each box until you have two letters the same. Record how many boxes you have drawn from. Replace the letters and repeat the experiment many times. If this were a game in which you could choose how many boxes from which to draw, and the prize for success in getting two letters the same was higher the *fewer* boxes you chose, then how many boxes would *you* ask for?

Experiments 2, 3 and 20 may all be analysed by the same underlying principle. Most pupils (and adults who have not already met the situation perhaps) feel that it is not very likely that two people in the class have the same birthday as there are so many more days in a year than children in a class. In fact, as we shall show later, the theoretical probability for a class of only 23 is over 50%, and for a class of 30 it is over 70%.

The arithmetic involved in the analysis of *Experiment 3* is simpler than in the other two, so we look at some results of that first. The following table shows a set of results as recorded by a group of children. They have referred to the months by numbers.

5	4	5	12	10	√	2	9	6	1	3		9	2	8	9	10	√	11	3	4	5	3	√	11	9	9	2	7	√
9	11	8	9	1	√	9	2	11	9	12	√	2	1	10	7	8		10	10	11	5	4	√	2	4	9	12	11	
8	1	12	7	8	√	6	4	3	9	9	√	2	5	5	8	6	√	3	10	6	8	6	√	5	11	12	6	6	√
12	7	6	8	9		9	4	1	12	2		9	9	12	1	8	√	4	9	10	8	5		11	4	12	6	1	
5	2	11	8	5	√	3	8	5	11	3	√	7	7	6	1	12	√	2	4	5	10	9		6	8	1	8	10	√
7	12	8	1	9		11	11	2	4	5	√	3	10	5	10	1	√	1	2	6	5	8		9	9	3	11	6	√
10	5	1	7	6		1	5	3	1	1	√	12	3	9	8	11		3	5	4	9	4	√	3	9	12	8	3	√
12	3	12	5	7	√	10	10	12	1	9	√	8	1	5	6	10		7	9	5	7	6	√	11	4	5	1	12	
2	5	6	8	10		11	9	12	4	1		2	5	5	5	10	√	8	10	12	7	7	√	12	4	7	2	10	
10	2	8	6	5		4	5	7	7	12	√	10	3	11	1	6		6	11	5	10	7		2	7	5	6	12	

Fig. 56

It is seen from the table that in 29 of the 50 sets of five people at least 2 had the same birth month giving a statistical probability from this experiment of 58%. The next diagram serves to represent the situation and we see that we have to investigate all the different ways in which the 'pin men can be put into the month boxes'.

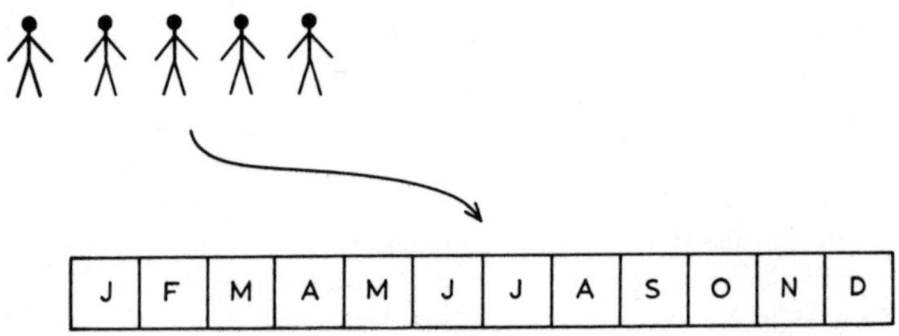

Fig. 57

One feels intuitively that there are quite a lot of ways of doing this! This first man can be put into any one of the 12 boxes and once he is placed the second man can be put into any one of the 12 boxes (one possibility is that he is in the same box). Therefore, there are 12 x 12 ways of placing just the first two. (Like the 144 outcomes of throwing two dice in the form of regular dodecahedra, 12 faced solids. The tree diagrams for each situation are the same, having 144 end points.) The third man can likewise be placed in 12 ways making 12 x 12 x 12 possible outcomes so far. The fourth and fifth men also have 12 ways each and hence the total number of ways of arranging the 5 men is 12 x 12 x 12 x 12 x 12 or, for short, 12^5.

The next task is to find out for how many of those 12^5 ways at least 2 men are in the same box. There are a number of approaches to this but the simplest is a method often used in such problems, namely to consider in how many cases *no two* are in the same box and then subtract this from the total number of ways –

(i.e. Total number of outcomes = (outcomes where two or more are in the same box) + (outcomes where no two are in the same box).)

If no two are to be in the same box then there are still 12 ways of placing the first man but only 11 ways for the second (making 12 x 11 so far), 10 ways for the third (12 x 11 x 10 so far), 9 for the fourth man and 8 for the fifth. Thus there are 12 x 11 x 10 x 9 x 8 outcomes in which no two are in the same box. It therefore follows that there are $12^5 - (12 \times 11 \times 10 \times 9 \times 8)$ ways in which at least two *are* in the same box. As the total number of ways is 12^5 altogether we say that the theoretical probability of our event is:

$$\frac{12^5 - (12 \times 11 \times 10 \times 9 \times 8)}{12^5}$$

Discussion of the various ways of simplifying this awkward fraction would be valuable – as would discussion of the alternatives to pencil and paper calculation. One obvious possibility is to change it to:

$$1 - \frac{(11 \times 10 \times 9 \times 8)}{(12 \times 12 \times 12 \times 12)}$$

This finally reduces to approximately 0·62 giving a theoretical probability which is very close to our practical results quoted earlier.

The analysis of these three experiments is probably the most difficult so far met and we do not suggest that all pupils should be expected to follow it – but a group may be able to develop the method with the aid of skilful questioning by the teacher. We would deplore any suggestion that thereafter pupils should be expected to memorise the results for future reproduction. They are included because there may be some pupils who may be interested in them, and to indicate some of the discussions about ways and means of computing which may arise from practical work.

For *Experiment 2* the principle is similar but this time there are more people and there are 365 boxes (what about leap year?). Suppose the class has 30 pupils. The total number of possible outcomes is 365^{30} – a very big number (usually *one* pupil will begin to try to calculate it!). Using the arguments of the previous experiment, the number of outcomes where *no two* are in the same box is (365 x 364 x 363 x . . . . . 338 x 337 x 336) and therefore the theoretical probability that at least two *are* in the same box (i.e. have the same birthday) is –

$$\frac{365^{30} - (365 \times 364 \times \ldots \times 338 \times 337 \times 336)}{365^{30}}$$

If we re-write it as:

$$1-\left(\frac{365}{365}\times\frac{364}{365}\times\frac{363}{365}\times\ldots\times\frac{338}{365}\times\frac{337}{365}\times\frac{336}{365}\right)$$

then pupils may be able to decide whether the probability increases or decreases as the number of people in the sample changes. Anyone who cares to work this out (it is not difficult with a school slide rule) will find it comes to 0·71 approximately.

The analysis of *Experiment 20* follows the same pattern. If we consider drawing letters from 12 boxes then there are 26^{12} possible outcomes (i.e. 26 from each of the 12 boxes – tree diagram with 26 branches at each junction). The number of possible outcomes with no two letters the same is given by 26 x 25 x 24 x 23 x 22 x 21 x 20 x 19 x 18 x 17 x 16 x 15. Therefore, the probability of 2 letters the same being drawn from 12 boxes is given by:

$$\frac{26^{12}-(26\times 25\times 24\ldots\times 16\times 15)}{26^{12}}$$

which is approximately 0·96 – a high chance. In the same way the probability that two letters the same come from only six boxes is

$$\frac{26^{6}-(26\times 25\times 24\times 23\times 22\times 21)}{26^{6}}$$

which is about 0·46. Similarly the probability for any number of boxes may be calculated and some of them are given in the next table.

Number of boxes	2	3	4	5	6	7	8	9	10	11	12
Probability of at least two letters the same	·04	·11	·21	·34	·46	·59	·70	·80	·86	·92	·96

Fig. 58

Again pupils would not be expected to work all this out for themselves but they might get some useful insight into the programming of a calculation and the use of tables or machines as a result of discussion and a co-operative effort.

The next table gives the results obtained by a group working on this experiment.

Number of boxes used before two letters the same were chosen . .	2	3	4	5	6	7	8	9	10	11	12	13 or more
Number of trials .	3	5	12	10	8	8	7	7	6	4	2	5

Fig. 59

The number of trials could easily be expressed as a percentage of the total number of trials and a percentage frequency graph drawn. It would not be easy to compare such a graph with the theoretical probabilities listed previously however because, for example, a probability of 0·46 for getting two letters the same out of six boxes would include cases where the duplicate letter had appeared after 2, 3, 4 or 5 boxes. In such cases it is appropriate to draw up a *cumulative frequency* table, i.e. one where in any total all previous frequencies are included, as shown below.

Maximum number of boxes	2	3	4	5	6	7	8	9	10	11	12	13 or more
Cumulative number of trials with two letters the same	3	8	20	30	38	46	53	60	66	70	72	77

Fig. 60

Fig. 61 shows the cumulative frequencies, expressed as percentages, compared with the theoretical probabilities.

6. If you roll a coin or disc 100 times down a chute, made from a piece of card folded, onto a board marked off in squares whose sides are twice the length of the diameter of the coin, how often would you expect the coin to land completely inside the square? Try it. If you were to double the length of the side of the square how often would you expect it then? Suppose you doubled the diameter of the coin instead?

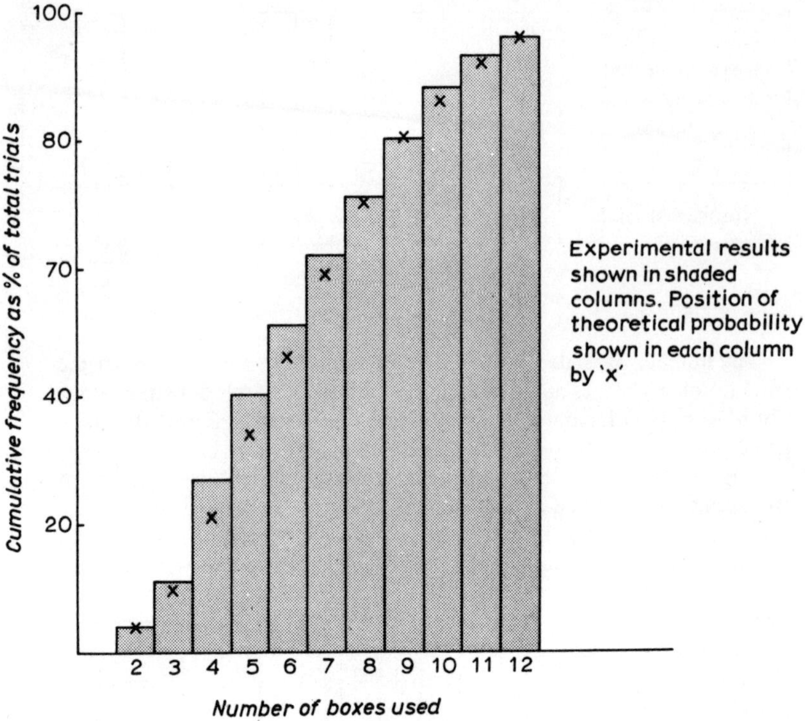

Fig. 61

The coin rolling activity of *Experiment 6* will be well known to children who visit fairgrounds. A chute made from folded card is quite adequate but a wooden wedge with a groove in it would be useful. Similarly drawing paper ruled or printed in squares is adequate but a more substantial board, made from a large sheet of hardboard perhaps, would make a suitable piece of maths room equipment which would be useful in a variety of situations. Pupils' initial *estimates* of the probability of the coin landing inside a square normally vary quite widely. For example the table gives the estimates from 80 pupils for 100 rolls of a coin on to a square grid of side twice the diameter of the coin.

'Number inside the square' predictions	0-9	10-19	20-29	30-39	40-49	50-59	60-69	70-79	80-89	90-100
Number of pupils	15	10	19	4	9	9	3	7	1	3

Fig. 62

The theoretical analysis in this experiment is a little different from any we have met before. The shaded region in this diagram shows where the centre of the coin can fall if the coin is to lie completely inside a square. What shape is the region?

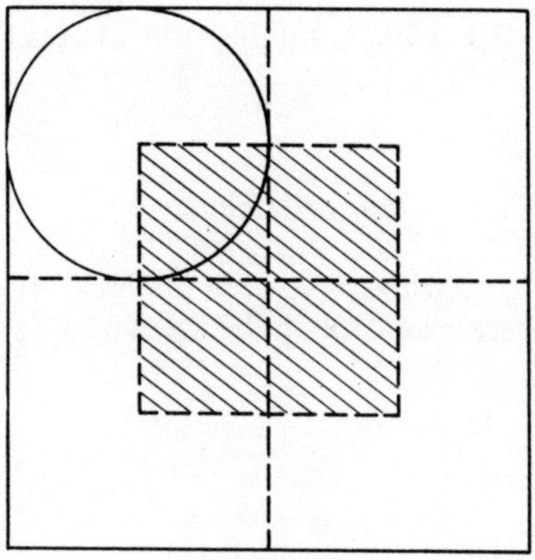

Fig. 63

If the diameter of the coin is d units then the areas of the squares of the grid are $4 \times d^2$, and the area of the shaded square is a quarter of this, d^2. Each point inside the grid squares is assumed to be an equally likely spot for the centre of the coin to fall, and therefore the probability is assumed to depend on the ratio of the shaded area to that of the whole square.

In this particular case the ratio is $d^2 : 4d^2$ or $1:4$ so the probability of a coin landing inside a square is said to be 25%. Notice that in the pupils' estimates quoted earlier, although the range was great, the most popular choice was the 20-29% interval. What would the ratio of the diameter of the coin and the length of the side of a grid square have to be to make the probability 50%? Similar experiments can be done on grids marked with other shapes such as hexagons or triangles.

5

Notes on The Classroom Activities (2)

Activities 22–32

Index to Chapter 5

As in Chapter 4, the activities have been grouped for analysis and this Index will enable specific activities to be located easily.

The last eleven activities in Chapter 2 are not *a priori* situations, in other words it is not possible to analyse all possible outcomes and hence calculate a theoretical probability. Any measure of probability is, therefore, a *statistical* one based on past events or on present conditions. Such conditions are usually assessed by some form of sampling. In using measures of probability so obtained to predict future events it is assumed that the future will follow the pattern (if any) of the past or, in other cases, that the whole 'population' behaves in the same way as the chosen sample. Sometimes such assumptions, although made, are not valid because other variables have been ignored or because the sample is somehow biased and not representative of the population it is meant to reflect. This may lead to false conclusions and ill-judged decisions which may have serious consequences in industry, commerce, science or government, etc.

The aim of this section of the activities is, therefore, to help pupils appreciate this way of estimating and using probability to make predictions using data *collected by themselves.* We specifically recommend that pupils should gather the data rather than be presented with information collected elsewhere, at least most of the time, as personal experience of data collection and tabulation can lead to more intelligent questioning and appreciation of facts presented in the press, on radio or on television. Even apart from this major consideration it is the experience of many teachers that pupils are likely to show more interest in information about themselves and their own environment than in pre-collected data which does not personally involve them.

22. Should a letter rack have 26 pigeon-holes, one for each letter of the alphabet, or would a different arrangement be better? Use a telephone directory to help you to decide on the best arrangement.

23. Should a word building game or a child's printing set have equal numbers of each letter of the alphabet? Use a book or a newspaper to estimate the frequency with which various letters of the alphabet appear in print.

Experiments 22 and 23 both involve frequency of letters of the alphabet, but in different situations. Probably most of us have been irritated by post-racks where some pigeon-holes regularly overflow and others are scarcely used. Those who lettered the rack may have *unconsciously* assumed that all the letters of the alphabet were equally likely as initial letters of surnames but experience shows that this is not so.

An analysis of the number of pages devoted to names beginning with the various letters in a county telephone directory is given below (less than half a page is shown as zero).

A	B	C	D	E	F	G	H	I	J	K	L	M
13	46	36	17	10	14	19	37	2	10	8	18	29

N	O	P	Q	R	S	T	U	V	W	X	Y	Z
9	5	24	1	20	40	17	1	3	33	0	2	0

Is it reasonable to assume that this sample is representative of the initial letters of the surnames of all the people in the county? Would the initial letter of the county name or any big towns have any bearing on this? Could a better source of names be used as a sample? Can you think of a situation where a sample made up entirely of telephone subscribers would almost certainly be a biased one? Discussion of how one might use the directory information, or any better source of data, to make a more efficient post-rack will be instructive. It is also interesting to compare data for different areas, South Wales, Scotland and the Midlands, for example.

Pupils often think of other situations where someone needs information about initial letters of names, for example manufacturers of initialled goods (handkerchiefs, cuff links) need such data to decide the proportions in which various letters should be used.

Experiment 23 also requires information about letter frequency, but from a different source as it concerns all the letters in words, and not just the initial ones of names. The table below gives the frequency of letters appearing in a sample paragraph of English prose.

A	B	C	D	E	F	G	H	I	J	K	L	M
44	9	13	13	61	10	9	27	37	3	6	20	16

N	O	P	Q	R	S	T	U	V	W	X	Y	Z
27	45	10	0	25	28	52	19	9	14	0	9	0

It is enlightening to try to crack a coded message where each letter of the alphabet has been replaced consistently by another letter. A comparison of letter frequency in the coded passage with that of a prose piece usually points to the way in which the most common letters have been coded and analysis of the 1, 2 and 3 letter words in the passage usually yields more. Pupils can try this in pairs. Another activity which usually yields very striking results is to ask each pupil to analyse a different paragraph for letter frequency and to list the 6 most frequent letters in order.

Unchecked results handed in by one class are given in Fig. 64. What discussion points would have arisen from them?

1st	2nd	3rd	4th	5th	6th
E	T	O	A	I	S
E	T	N	H	A	O
E	T	N	H	I	O
E	T	N	I	O	R
E	O	A	R	I	L
E	A	T	S	R	I
T	A	W	I	S	R
E	T	R	A	I	S
E	O	T	N	R	S
E	A	S	I	T	N
E	A	T	R	S	L
E	T	S	A	N	H
E	T	O	A	R	S
E	A	S	T	I	L
E	N	T	A	O	R
E	T	O	H	I	S
E	T	S	R	A	H
E	T	N	A	D	S
A	E	T	O	I	N
A	T	O	E	N	R

Fig. 64

24. Collect information about the number of brothers and sisters which people in this class have. If you choose someone from the class at random, is he or she likely to have at least one brother or at least one sister?

Information about sizes of families, used in *Experiment 24*, can lead to interesting sociological discussion, but teachers must exercise judgement about its suitability in their particular school. The tabulation given in Fig. 65 relates to 108 children in the 14–15 year age range. The numbers in the squares show the total number of families having the number of boys and girls indicated on the axes. Thus the commonest family make-up in this sample was 1 boy and 1 girl. The shaded squares represent families with 2 children.

What is the pattern for the families with 3 children?

In this sample are there more families with 2 children or more with 3 children?

What information is obtained by summing the rows and by summing the columns?

Frequency graphs could be drawn from these sums.

Where are the families with the same number of boys and girls?

Number of girls \ Number of boys	0	1	2	3	4	5
7					1	
6						
5		1	1			
4	1		1			1
3	4	1	3		1	
2	6	14	9	4	1	
1	5	17	8	4	1	
0		4	13	6	1	

Fig. 65

Where are the families with more boys than girls and those with more girls than boys?

We see that 35 families have one girl and 37 families have 1 boy. Does this mean that 35 + 37 families have 1 boy or 1 girl? The Venn diagram for the situation looks like this:

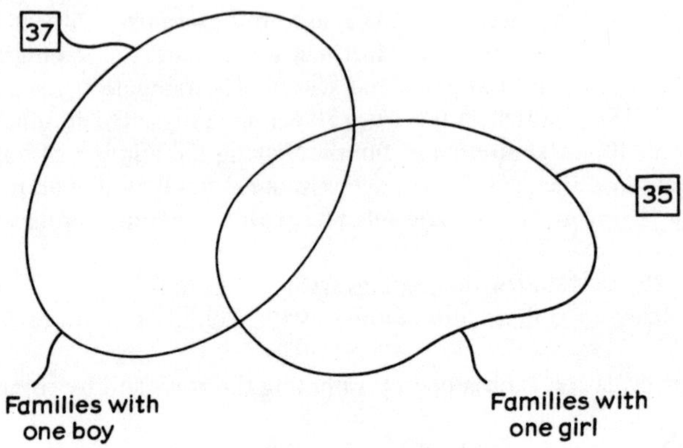

Fig. 66

Fig. 67

Why are there no families in the (0, 0) square of Fig. 65? (This question sometimes provides a surprising variety of answers.)

This situation offers a good opportunity to make a 3-dimensional frequency graph by piling cubes (wooden, card or paper) on the squares to represent the appropriate number of families (Fig. 67).

31. On any one Saturday in the season about how many goals are likely to be scored altogether by the 1st, 2nd, 3rd and 4th divisions of the Football League? What is likely to be the commonest number of goals scored by any one team? Are more goals likely to be scored by 4th division teams than by 1st division teams? Use past results to help you to find reasonable answers to these questions.

The football league results used in *Experiment 31* can be tabulated and graphed in ways similar to the information about family sizes. Fig. 68 shows the matches of one week classified and counted according to the number of Home and Away goals scored.

Away goals						
5	0	1	0	0	0	0
4	0	0	0	0	0	0
3	0	0	0	0	0	0
2	3	1	2	0	0	1
1	3	8	4	0	0	1
0	8	3	2	3	0	2
	0	1	2	3	4	5
	Home goals					

Fig. 68

Which figures in the table represent drawn matches?

Which represent home wins and which away wins?

Some pupils may be interested in keeping week by week records and using them to try to predict future results. Teachers who are concerned about gambling may find that this discourages rather than encourages it!

25.	Should shoe shops stock equal numbers of all sizes of shoes so that they can be fairly sure of having something to fit any customer? Record the shoe size and foot length (without shoes) of as many people as possible. Is there any relationship between these two measures?

26.	Collect as much information as possible about people's foot-length and foot-width (without shoes). Why would information of this type be important to a shoe manufacturer?

27.	Are boys taller and heavier than girls of the same age? Is it likely to be true that for boys armspan is usually more than height and for girls armspan and height are usually the same? Measure the height (without shoes), the weight and the arm-span of as many people about your age as possible.

A number of important ideas can be explored through activities involving human measurement and *Experiments 25, 26 and 27* are just a selection of the many possibilities. Experiments 25 and 26 are about shoe sizes and foot measurement and they invariably produce very interesting results. Some pupils, especially older girls, have been known to develop this work into a major sociological project.

Two of the four frequency graphs, (Fig. 69 a–d), illustrate the sizes of shoes worn by a sample of just over 100 pupils in a certain one year age range. The other two illustrate the distribution of foot lengths, to the nearest centimetre, of the same sample. Pupils may make a number of observations about such graphs. They may say, for example that 'boys seem to have bigger feet than girls'. Do they mean that any one boy chosen will have bigger feet than any one girl? If not what *do* they mean? *The idea of an 'average' instinctively comes in.* They may also observe (by comparing the size graphs and the length graphs, if they did not notice this when they recorded the information) that shoe size and foot length do not seem to be as related as they thought they would be. (Look at the lower ends of the girls' graphs for example!)

This becomes more strikingly obvious if one compares the individual shoe size and foot length using a 'scattergram'. Such a graph is shown in Fig. 70. The number of dots in each square represents the number of children having a particular shoe size and foot length. The graph is similar

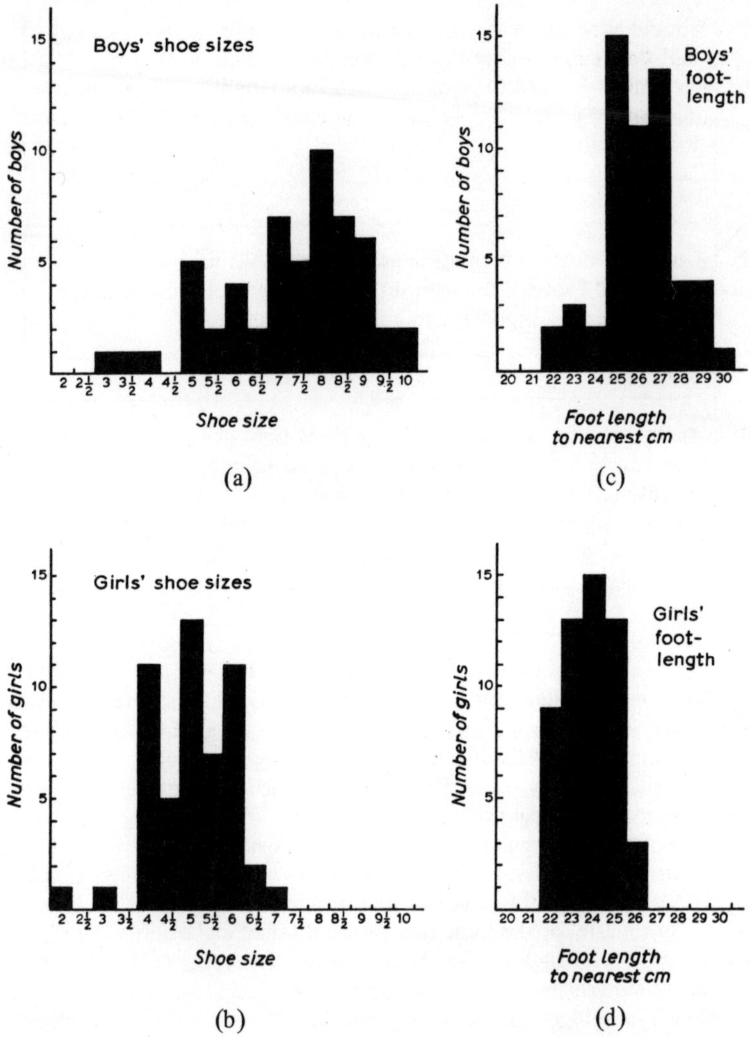

Fig. 69

to the tabulations of family sizes and football results met earlier in this chapter. It is also a special form of relation graph (see Chapter 3) where the axes are divided into block intervals instead of being scaled continuously because the lengths have been adjusted to the nearest centimetre and shoe sizes are not continuous, eg. there are no sizes *between* 6 and $6\frac{1}{2}$, etc.

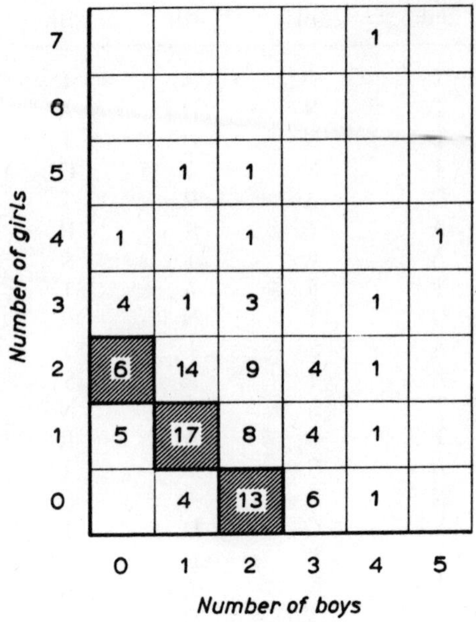

6

Arrangement and Choice

Competitions, classification and number of ways – simple permutation: choice and permutation (^nP_r): combination (^nC_r): powers (r^n).

In all the *a priori* situations analysed in Chapter 4 we were concerned with the possible outcomes of any trial and, in particular, with the *number* of outcomes and the ways in which they could be classified.

More work of this type can come from the entry forms for the competitions which appear on packets and tins, in domestic magazines, shopping guides and newspapers. Most children would easily be able to find several at home.

Some of the competitions they produce will undoubtedly be of the straightforward 'put in order' type. Frequently the competitor is invited to put, say, six statements or pictures in a preferred order according to a described situation. Diagrammatically we have:

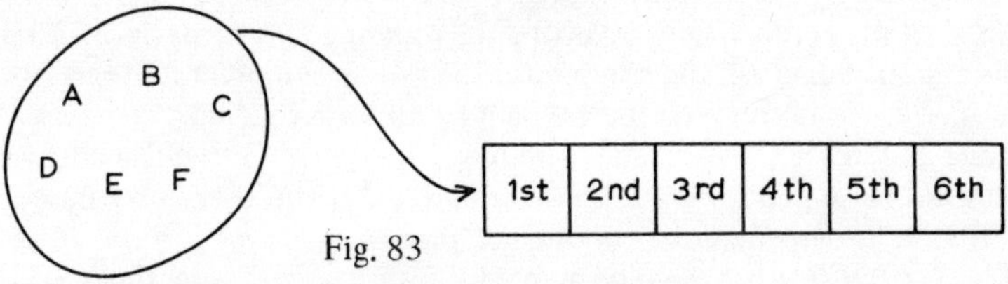

Fig. 83

A tree-diagram, Fig. 84, shows the possibilities

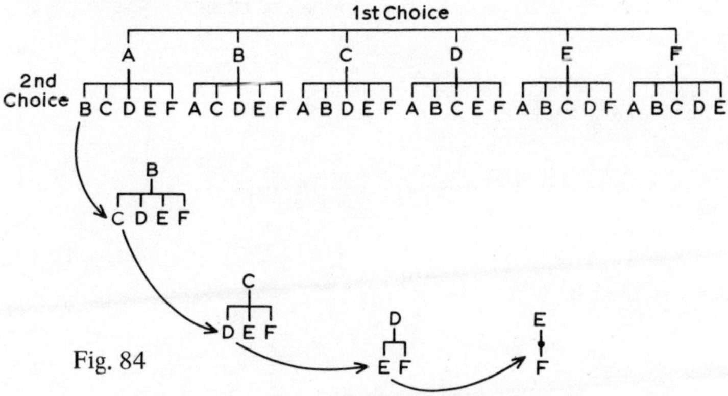

Fig. 84

There are 6 ways of making the first choice and *for each one of those 6 ways* there are 5 ways of making the second choice, hence there are 30 ways (6 x 5) of choosing the first and second, as indicated by the 30 branches of the tree diagram at that point.

Similarly *for each of these 30 ways* there are four ways of choosing the third item – hence the tree has 6 x 5 x 4 branches after 3 choices. So the pattern continues (space does not permit the whole diagram to appear). After the fourth choice there are (6 x 5 x 4 x 3) branches, after the fifth there are (6 x 5 x 4 x 3 x 2) and after the sixth choice (which is really no choice at all of course) there are (6 x 5 x 4 x 3 x 2 x 1) branches.

This expression, (6 x 5 x 4 x 3 x 2 x 1), is known as 'six factorial' or 'factorial six' and is commonly written in symbolic form as 6! (Sometimes $\underline{|6}$.) An arrangement of elements in some order or another is technically known as a *permutation*, hence 6! is the total number of different permutations of six elements. This number is in fact only 720 (we say 'only' because, for example, 10! approaches 4 million). One can imagine therefore the feasibility of entering all possible permutations in a competition involving the ordering of only six items – it may even seem to be worth eating 720 tins of . . . Beans in order to send a label with each line! Hopes are usually dashed however by the necessity to write a slogan, 'I have . . . on my cornflakes because . . . ' and the chance of pairing the winning slogan with the one correct line seems remote.

Not all the competitions involving the ordering of 6 items have 720 possible entries. Some ask the competitor to choose, and order, 6 items from a list of, say, 10. In this case there would be 10 ways of choosing the 1st, 9 for the 2nd, 8 for the 3rd, and so on down to 5 ways for the last choice giving a total of (10 x 9 x 8 x 7 x 6 x 5) i.e. the first 6 terms of 10!

Thus it is said that the number of *permutations of 6 from 10* is (10 x 9 x 8 x 7 x 6 x 5). The symbolic form is $^{10}P_6$ (sometimes $_{10}P_6$). Once one has built up a table of values of various factorials it is sometimes useful to be able to express such a number as 10 x 9 x 8 x 7 x 6 x 5 in factorial form.

This is easily done by writing it as

$$\frac{10 \times 9 \times 8 \times 7 \times 6 \times 5 \times 4 \times 3 \times 2 \times 1}{4 \times 3 \times 2 \times 1}$$

which can then be written as $\dfrac{10!}{4!}$

Similarly $^7P_4$ is 7 x 6 x 5 x 4 or $\dfrac{7 \times 6 \times 5 \times 4 \times 3 \times 2 \times 1}{3 \times 2 \times 1}$ or $\dfrac{7!}{3!}$

and $^9P_3$ is $9 \times 8 \times 7$ or $\dfrac{9 \times 8 \times 7 \times 6 \times 5 \times 4 \times 3 \times 2 \times 1}{6 \times 5 \times 4 \times 3 \times 2 \times 1}$ or $\dfrac{9!}{6!}$

hence in general form, $^nP_r = \dfrac{n!}{(n-r)!}$

Competitions of the ^nP_r type seem to be the most common, but other types of competition include some football pools. Suppose we consider the 'Treble Chance' (why that title?). In this one does not have to put anything in order, but one has to choose eight matches from the fifty listed, as potential 'draws'. This particular example, which is almost bound to arise if pupils are invited to supply the material, is perhaps best approached through an activity involving smaller numbers where the structure will be clearer.

Using five colours of sticky paper a group can display the $^5P_3$ (i.e. 60) ways of choosing, and ordering, three of them, e.g. if the colours are Red (R), Green (G), Blue (B), Yellow (Y), and White (W), a colour chart might be made thus:

RGB	GBR	BRG	BGR	RBG	GRB
RGY	GYR	YRG	YGR	RYG	GRY
RGW	. . .	. . .	. . .	. . .	. . .
GBY					
GBW					
BYW					
RBY					
RBW					
RYW					
GYW	. . .	. . .	. . .	. . .	. . .

Fig. 85

Clearly there is an opportunity to discuss system and pattern here and to count the number of times each colour appears in each row and column. We then notice that any one colour combination appears in the tabulation in all its possible orders (in this case in 3! orders as there are three colours in each combination) i.e. RGB, GBR, BRG, BGR, RBG, GRB.

As we are interested only in choosing (but not putting in order) a combination of 3 from 5 we have, in $^5P_3$, *6 times as many as we want.* Hence there are not 60 ways but $\frac{60}{6}$ (i.e. 10) ways of choosing 3 from 5. This number is called the number of *combinations* of 3 from 5 and is written as $^5C_3$. It is computed by dividing $^5P_3$ by 3! i.e. by the number of ways the elements in the combinations may be permuted amongst themselves.

In general $^nC_r = \frac{^nP_r}{r!}$ or $\frac{n!}{(n-r)!r!}$ (see page 93).

If we apply this idea to the treble chance pools we have to compute $^{50}C_8$ which is $\frac{50 \times 49 \times 48 \times 47 \times 46 \times 45 \times 44 \times 43}{8 \times 7 \times 6 \times 5 \times 4 \times 3 \times 2 \times 1}$

i.e. $\frac{\text{Number of permutations of 8 from 50}}{\text{Number of permutations of 8 themselves}}$

Probably at least one pupil will be motivated strongly enough to attempt the calculation (the answer is in the region of 540 million) and hence to compute the cost of a foolproof entry.

We have already met, in the analyses of Chapter 4, situations with coins and dice which involve combinations and in particular, on page 53, the beginning of Pascal's Triangle was written in ^nC_r form. Those who are familiar with the so-called *triangular numbers* 3, 6, 10 . . . which appear in Pascal's Triangle might consider why these turn out to be the combinations ($^3C_1$, $^4C_2$, $^5C_3$ . . .) or ($^3C_2$, $^4C_2$, $^5C_2$ . . .) and, if 1 is a triangular number which could be written as $^2C_0$, what value should we give to 0! (factorial zero)?

A fourth type of competition is common and can again arise from football pools if we consider the number of ways of forecasting, say, 6 results.

Diagrammatically we have:

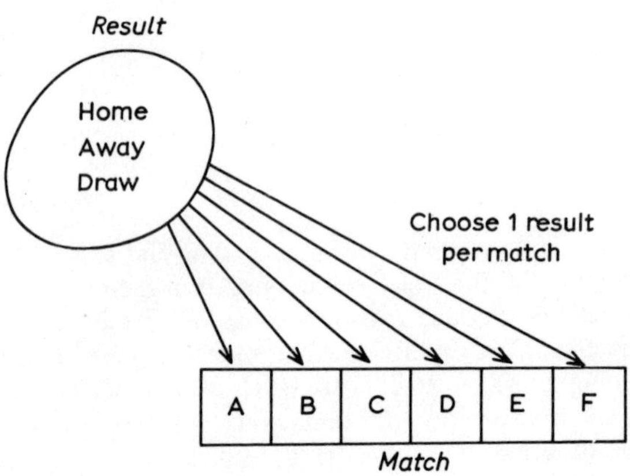

Fig. 86

There are 3 choices for match A, 3 again for match B, 3 for C . . . and the tree diagram has three branches at each junction . . . . .

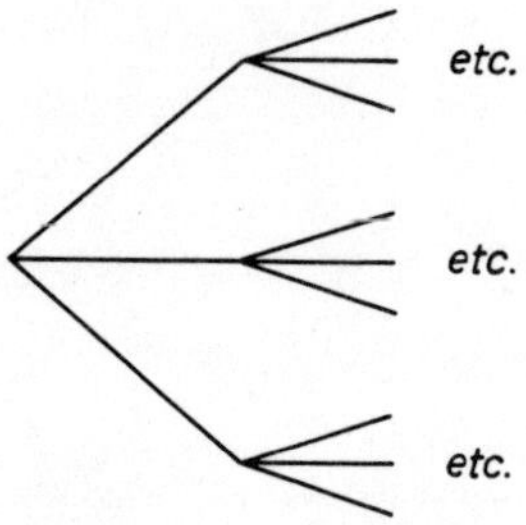

..... giving a total of
3 x 3 x 3 x 3 x 3 x 3
branches which is written, for short, as 3^6.
(In speech 'three to the power of six' or simply 'three to the sixth'.)

Fig. 87

We have met this pattern before in the total number of outcomes of tossing coins or throwing dice. In general the number of outcomes from n items each of which can occur in r ways is r^n. (Compare for n coins the result 2^n and for n cubical dice the result 6^n.)

Experience shows that most of the competitions pupils will produce will fit into one of the categories we have discussed, or at least be related to them. We strongly recommend, for reasons concerned with motivation, that the pupils supply the material for consideration. Apart from the general knowledge it promotes, a study of competitions can obviously be useful from several points of view. Some calculation (with a degree of motivation) will occur, together with an opportunity to discuss ways of simplifying it, the chances of winning, the cost of a foolproof winning entry, the economics of the situation for the competitor and the promoter. General formulae may emerge, there will be an opportunity to look at the structure of the situation and hence to classify it and, if possible permutations or combinations are listed, there is a great deal of systematic thinking and pattern making to be done.

7

Questionnaires and Information Storage

What sort of questions? – Topics for surveys – official forms and tests – punched card systems: notched cards, recording methods, coincidence cards, cards for electronic sorting.

'If you want a job well done do it yourself.' This is as good a maxim in data collection as in many other things. Sample data from a population (be it people or things) are likely to be best collected by the person who is to analyse them, insofar as he knows just what information he wants and in what form he wants it. The experiments we have suggested all involve the pupil in gathering his own data.

Such a course is not always possible in real life. There cannot always be person to person contact between the information collector and every element of his population. Sometimes it is not even desirable as it may introduce 'experimenter bias'. In such cases a questionnaire is often used.

We think that the opportunity to make up and test a questionnaire is an important part of any pupil's total experience of statistical work. Teachers will be well aware from their own experience in writing worksheets, exam papers, etc. how unexpected interpretations can be put on seemingly (to the writer) clear questions.

Even apparently simple points on a questionnaire, like 'age' or 'colour of eyes', can produce much uncertainty. 'Do you want my age last birthday or do you want the months as well?' Such a question may well be asked if both parties are present, but in the absence of the analyser the subject must make up his own mind and it may matter that different subjects make different decisions. It would be better to ask for 'date of birth' and, provided of course that the questionnaire is dated also, the analyser can be as accurate as he wishes to be in recording the age. Similarly a test run on eye-colour, especially in a class of girls, may produce an astonishing range and the questioner would be better advised to say 'underline the colour which is nearest to that of your own eyes' – followed by a list – although even this course is open to criticism. In fact any question where the subject is asked to classify himself in some way, without being given a list of classifications from which to choose, is almost bound to fail.

It seems pointless however to give pupils a long list of do's and don'ts before they begin to compile a questionnaire themselves. They are likely to learn far more effectively if they see the difficulties in action rather than if they are steered round them. We suggest therefore that a test run on a small number of subjects before the questionnaire is finally duplicated and circulated is an appropriate course.

There is no shortage of suitable topics for surveys of fact or opinion which can be undertaken in most schools by questionnaire. We list a few which some teachers have found successful:

Holidays (where, with whom, accommodation, duration . . .), leisure (clubs, dances, T.V., sport, pocket-money, papers and magazines), work (future, part-time jobs now . . .), opinions (politics, uniform, school meals, rules . . .) . . . Pupils will suggest many more. Teachers need, of course, to judge the suitability of any one topic in relation to their particular school environment and neighbourhood.

As well as making up and using questionnaires it is also valuable for pupils to fill them in. We are something of a nation of form-fillers and all our pupils, no matter how unintelligent, will meet officialdom in this way. It would be particularly useful for pupils to examine applications for passports and forms for driving licences and the insuring and licensing of vehicles. The completion of these is not always straightforward even to adults and most people come across them at one time or another. They may also be interested in tests of personality or attitude which employ a number of different techniques to obtain meaningful answers. Many of these are printed for self-testing in magazines or papers and many more are used in industry, commerce and the social services. Opinion polls and market research questionnaires are also fruitful.

Teachers may be able to obtain copies of the forms, tests and questionnaires mentioned from Government or Council departments, University Departments of Education or of Psychology, the Armed Forces, Public Relations Organisations, Personnel Agencies, local industry, national newspapers, research institutes or agencies, etc.

Information Storage and Retrieval

Information obtained from questionnaires in industry, research or the social sciences if often processed and stored in some way until required. Frequently the processing consists of transferring the information to a card, or paper tape, by selective punching. It is then stored and used in a manual or electronic machine. The machine is used for subsequent sorting or retrieval of the information. The electronic types come under the heading of *computers* although, at least in this respect, they sort (and often print out) rather than compute.

We cannot give an exhaustive list of the various methods used but we discuss some which are simple enough for pupils to grasp and which may serve as a basis for the understanding of other systems which they may be able to see locally or which they may meet at work later on.

The Personnel Department of a small factory may well use an 'edge-punched' or 'notched' card to store information about employees. Such a card, unprocessed, might look rather like this:

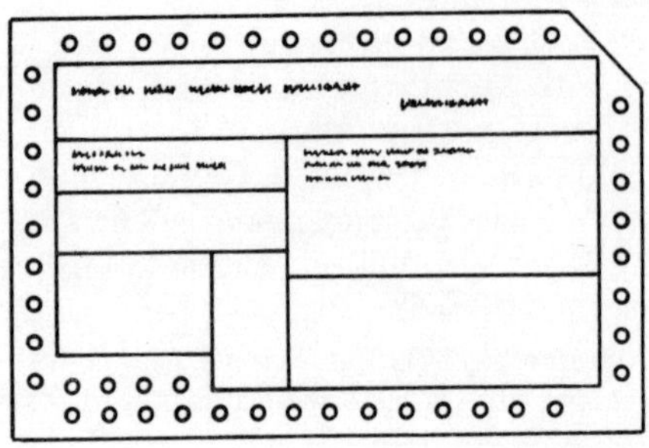

Fig. 88

In general the holes are in a single line round the edge of the card but sometimes a double line of holes appears. Each hole can be used to record a piece of information and this is done by turning the hole into a 'notch'. In a very small installation this might be done with a small hand-tool and in a larger installation a manually operated machine rather like a typewriter might be used.

Fig. 89 shows two adjacent holes labelled M (for male) and F (for female). The appropriate hole is notched for each employee, thus (b) shows the record for a female and (c) for a male.

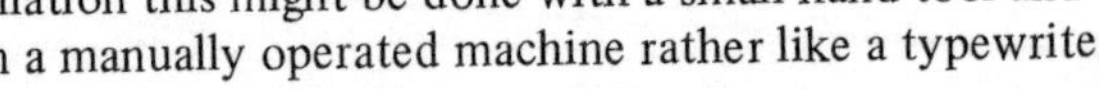

(a) (b) (c)

Fig. 89

It would of course be more economical of holes to use just one and notch it for females, say, and leave it for males but to do so would mean that an unprocessed card, by being unnotched, was storing information. When two holes are used one *has* to be notched and this serves as a check that the information has been recorded; such a check could not operate if only one hole were to be used.

Information about department, qualifications, family, etc. can all be transferred to specific holes round the card. A number of ingenious systems are used. Numbers for example may be expressed in the binary system (see *Number Appreciation*) where they will have only 0 and 1 as digits and the

1's can then be recorded as notches. Fig. 90 shows the number 524_{ten} expressed as 1000001100_{two} (for easy interpretation as a base ten number the holes are labelled with their base ten value).

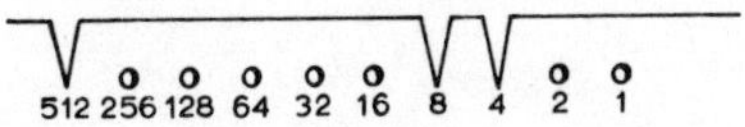

Fig. 90

The arithmetic involved in this denary-binary conversion however makes it desirable to have a coding method which uses the familiar base ten number directly. One such method is shown in Fig. 91 where 524 is again recorded.

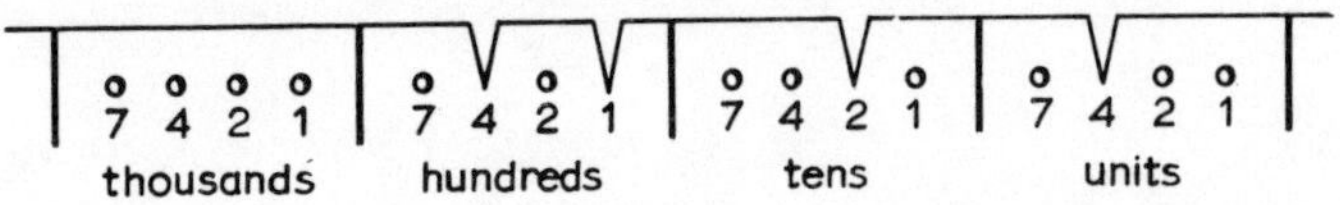

Fig. 91

The digits are recorded by notching a hole, or a combination of holes, in the appropriate section. It will be seen that by using a 7 for the left hand hole in each section a maximum of *two* notches are required to record any digit. Pupils might investigate the relative number of holes needed to record up to 999, 9999, etc. in each system and the relative number of notches required to record various specific numbers either way.

Another coding system uses the double bank of holes. For example the months of the year would take a line of 12 holes but can be recorded on a double bank of only 8 holes. Fig. 92 (a) illustrates this and (b) and (c) show how January and February respectively would be recorded.

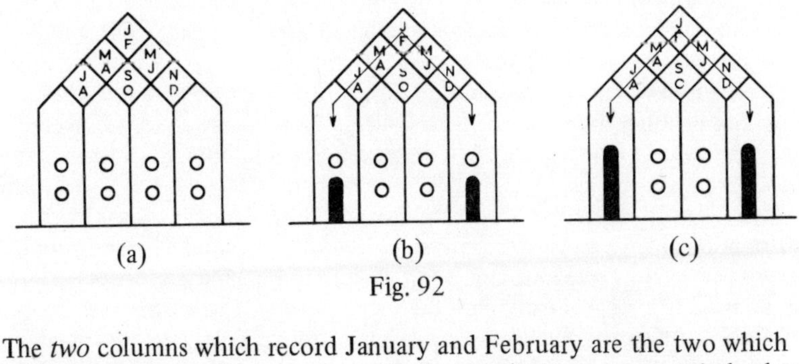

(a) (b) (c)

Fig. 92

The *two* columns which record January and February are the two which are reached by travelling out from the January/February square in both directions. January is then recorded by notching the *first* holes and February by notching *both* holes in the two columns.

The six squares each pick out two pairs of holes (${}^4C_2$, the number of ways of choosing 2 from 4, is 6). If another pair of holes were added then we would have ${}^5C_2$ squares, i.e. 10 squares.

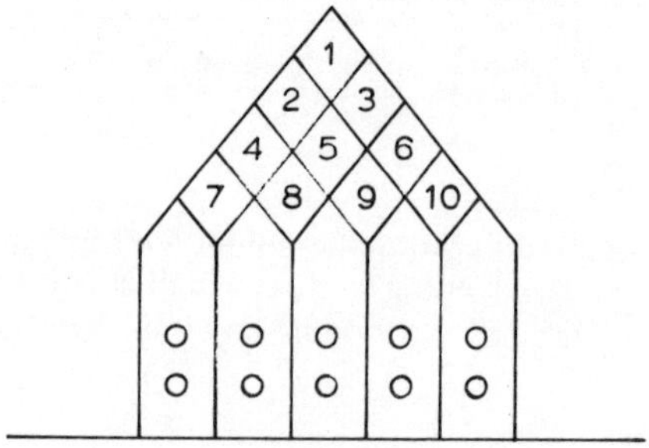

Fig. 93

Six pairs of holes would yield ${}^6C_2$ or 15 squares in which 30 pieces of information could be stored. We notice the triangular numbers 6, 10, 15 ... appearing again. The 26 letters of the alphabet can be stored in a 6 column double bank of holes and initial letters of surnames are sometimes thus recorded.

Simple punched cards are now often used in the mathematics classes of secondary schools. They can be purchased by the thousand through educational suppliers and sometimes through the L.E.A. stationery department as library index cards punched to a desired pattern. Pupils can thus consider how best to transfer information they have collected to a home-made punched card system. In one large comprehensive school, several years ago, all the entry subjects of some 100 5th form examination candidates were recorded, by the candidates themselves, on individual punched cards. The school secretary, who had to type 3 copies of the candidate list for each individual subject, simply stacked and aligned the cards and, using a milk straw inserted in the appropriate subject hole, lifted the pack and the cards for the candidates in that subject fell out. In this way she saved much searching through these massive entry sheets! Pupils may think of several practical ways in which a punched card system could be used in their own school.

Variations in the milk straw technique are the basis of retrieval of information from a small storage system like this. A metal case and prong are usually used to pick out any set of cards required and successive sortings can be used to find, for example, 'the men in Dept. X who have 3 or more children' quite easily. Such simple mechanisation is clearly an advantage over the longer task of scrutinising each employee's individual record.

A number of books for pupils which we list in Appendix 1 describe worthwhile work using punched cards.

Another type of punched card information store is the 'coincidence card'. A typical card might look like this:

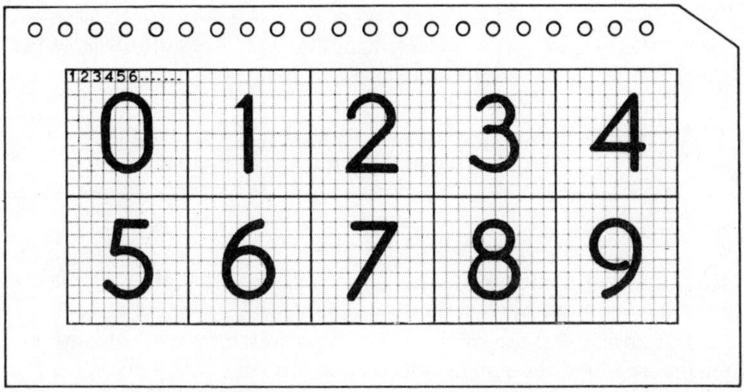

Fig. 94

There are 1,000 small numbered squares on this particular card, 100 of them in each of the large squares numbered 0 to 9. All the elements in the population (maximum 1,000) have a reference number and hence can be represented by a specific small square on the card. For example the 'population' might be all the books in a library, which, being numbered from 1 upwards, each have a place on every card.

Any one card is used to record one single classification of the elements of the population. For example we may wish to classify the books into those which are novels and those which are not. One card is then headed 'novels' and hole is punched out from each of the squares which represents such a book. This is done using a sort of long-arm punch or other mechanical device. If the completed card is then held up to the light or, as is often done, is held over a glass plate, then light shines through the holes that are punched in the squares which represent the novels. If another card were to be taken, recording, for example, those books which are about Australia then, when one card is superimposed upon the other the light shines through the holes representing novels about Australia – hence the name *coincidence cards.*

When a new element is added to the population (e.g. a new book to the library) then all the cards appropriate to that element are sorted out (often by an edge-notching system, hence the edge holes along the top of the card) and may be punched simultaneously through the square allotted to the new element.

It is interesting to note the relationship between the use of notched cards and coincidence cards as shown in Fig. 95.

	Each card represents	Each hole represents
NOTCHED CARD	1 element in the population	1 classification of the population
COINCIDENCE CARD	1 classification of the population	1 element in the population

Fig. 95

Thus we see that the two systems are reversals or 'duals' of each other. The very commonly used *two-way entry table* (of which a school register is a particular example) is a combination of both systems with the elements along one edge and the classifications along the other. Any one row of the example shown in Fig. 96 is equivalent to a notched card and any one column does the job of a coincidence card. For example we may take a notched card for Mr. A and record him as belonging to classifications a, c, d and f. Alternatively we could take a coincidence card for classification 'a' and punch through the spaces allocated to Messrs. A, B, D and E.

Set

Population	*a*	*b*	*c*	*d*	*e*	*f*
Mr. A	●		●	●		●
Mr. B	●	●		●	●	
Mr. C		●	●	●		●
Mr. D	●			●	●	●
Mr. E	●	●	●			●

Fig. 96

A type of coincidence card often turns up in number puzzles where it is frequently referred to as a 'window' card. Pupils can make their own coincidence cards from squared paper. Fig. 97 shows a design for a card to hold information about a population consisting of a class of 25 children.

Fig. 98(a) indicates that the children numbered 3, 5, 6, 8, 13, 14, 16, 18, 20, 22, 24, (i.e. the *missing* ones) are boys and Fig. 98(b) shows that those numbered 7, 13, 16, 20, 23 are in Tudor House.

When superimposed (Fig. 99) we see that 13, 16, 20 are the boys of Tudor House.

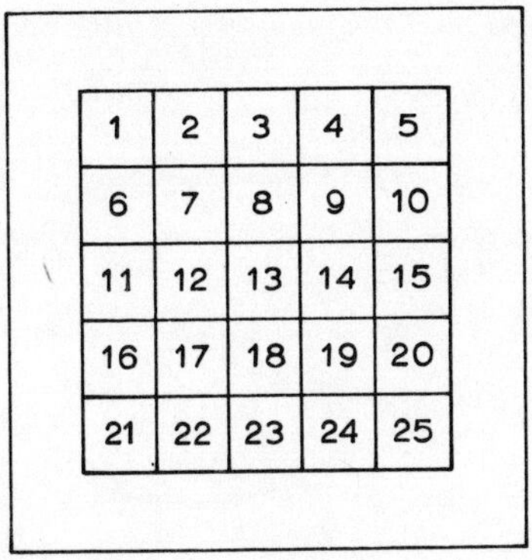

Fig. 97

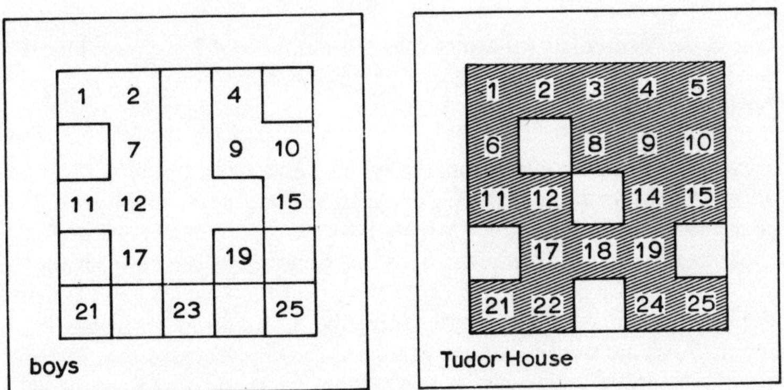

(a) (b)

Fig. 98

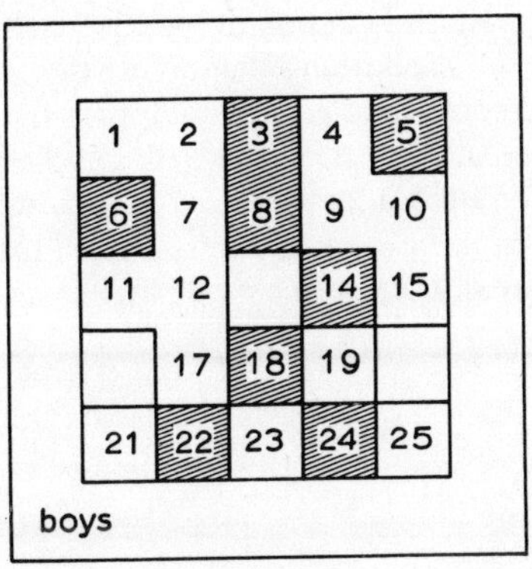

Fig. 99

The situation can also be illustrated by a typical Venn diagram, thus:

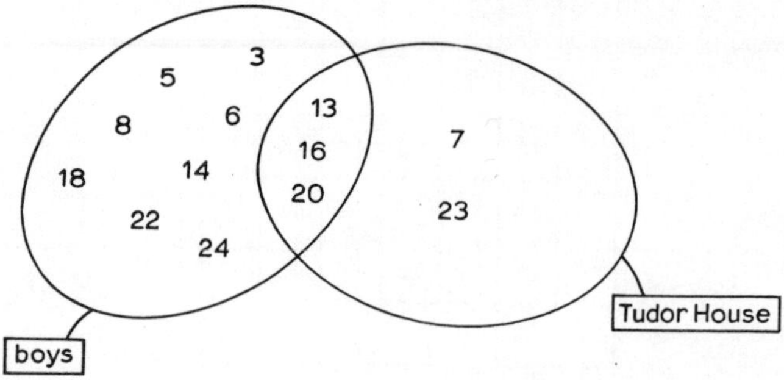

Fig. 100

Where in the Venn diagram are the shaded numbers of Fig. 99? Who are they? Does it make any difference if we reverse the order of the superimposition in Fig. 99?

Teachers will also come across the type of punched card which is used in an electronic sorting machine. The electronic machine would probably appear in a big installation, but the card is used for the same sort of purpose as the notched card. The holes are punched in the body of the card, but it is not a coincidence card for one card represents one element in the population and the holes represent the information about that element. Holes are used to act as 'windows' for the electronic impulses by which the cards are sorted just as the notches are used to let the 'needle' get by with a notched card.

There is certainly no lack of opportunity for pupils to see information storage and retrieval systems in action in local industry and commerce if the teacher is willing to make arrangements for visits. We recommend however that they do some simple classroom work first so that they can better appreciate some of what they see. With rapid progress and expansion in the information field it may even be possible to acquire small items of equipment which firms have outgrown – a small manual punched card system would be of great interest in a mathematics room.

8

Sampling

Transferring information to another population – relation between sample and population: matching, homogeneous populations, randomness, sample size – summary – classroom experiments.

In the last chapter we were concerned with the gathering, storing and analysis of information from certain populations. The purpose of those data collections was to record and study some features of those populations themselves.

However, it frequently happens in life that we obtain information about *one* population and use it to make statements (or assumptions) about *another* population, often a larger one including the original. Such a step is full of pitfalls and perhaps it is on account of the ways in which people have used this idea from time to time that one sometimes hears ridiculous statements like 'You can prove anything with statistics'.

We all, probably unconsciously, transfer information from one population to another in the course of our everyday lives. If someone says, 'I would never have thought you were a maths. teacher' they base their statement on assumptions they have made about all mathematics teachers on the basis of data they have collected from those they have known! Other such examples are plentiful whenever we classify our fellow human beings, and they make an interesting sociological study.

Reports of opinion polls, T.V. viewing figures, etc., are easily obtained from the national press and teachers may find it helpful to introduce the ideas behind this aspect of sampling by using such examples, found by the children themselves. Ask such questions as:

'What is the population about which the assumptions are being made?'

'Are *you* in this population?'

'Were *you* asked for information?'

'Who do you think *was* asked?'

If the teacher can obtain authentic details about how one such survey is really done then so much the better – it may even be possible to arrange for a researcher to visit the school and speak to the pupils.

From the discussions will emerge the fact that only a relatively small sample is asked and yet statements are made (or if not made are implied) about the larger population. Is the sample even always part of the population about which assumptions are made? *Not* always: two cases commonly arise which may be diagrammatically represented in Fig. 101.

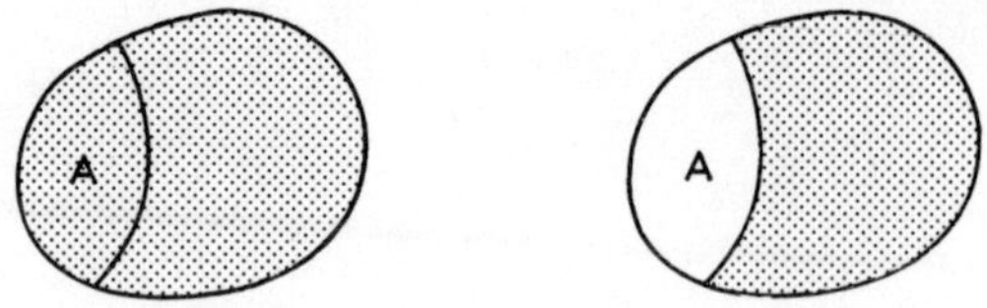

(a) (b)

Fig. 101

The shaded area represents the population about which assumptions are being made based on the information from the 'sample' of the population, shown as a subset, A.

Case (a) is the most common but (b) can arise for example if a sample from a batch of manufactured goods is tested to destruction in some way, then any assumptions made from this test about the proportion of sub-standard items, length of life, etc. can only be applicable to *those which are left*.

Will all samples produce acceptable results? It depends how we define acceptable of course. What steps can be taken to see that the assumptions are as valid as possible? Ideas of 'matching' the sample with the population will begin to be aired. Would it be reasonable to try to make assumptions about the number of houses with central heating in an area by telephoning a sample of householders? Clearly it would not. Common sense tells us that this may introduce a significant bias. *Common sense?* No. Just our own previous, but unconscious, sampling experiences which may make us assume that a sample of telephone subscribers is likely to yield a higher proportion of homes with central heating than is the case amongst houses as a whole. Our previous experiences can thus help us to eliminate potential bias, but it follows that, where the experimenter has no previous relevant experience, bias unrecognised by him may, and often does, creep in.

Many books for teachers and children on this topic give excellent examples for discussion and we refer the reader to the book list in Appendix 1.

We need to have a reason to believe that the proportions in which an event appears in the population and in the sample are likely to be the same. An experiment with a sampling bottle illustrates this well. Two bottles are taken, both containing the same fairly large number of coloured beads. The same quantity of white beads as coloured ones is introduced into each bottle and *one* bottle is well shaken. Several samples of a fixed size are then taken into the tubes by tipping, not shaking the bottles. Although the proportion of white beads is the same in each case the results of the sampling are likely to show marked differences. This occurs because the sample was not taken from a well mixed, or *homogeneous*, population.

This is not to say, however, that acceptable results *cannot* be obtained from a non-homogeneous population; if groupings within such a population can be identified then the sample can be made 'representative' of it by reflecting the groupings. (This is almost always done in social surveys.) The subjects of randomness and representative sampling are vast and teachers are advised to consult more specialist literature than this to study them. Suffice it to say that we frequently describe events as random when they are in fact not so. (Try *choosing* two of your class at *random* to collect some equipment from another room and you will probably see what we mean!) In general randomness is something which human beings find extremely difficult to achieve without mechanical or electrical aids and even then unexpected bias can arise. A study of actual methods used in science and industry can be instructive.

Another pitfall concerns the size of the sample in relation to the population. There is a great deal of mathematical theory in this connection but at this level we should again concern ourselves with intuitive ideas. It may well be worth discussing the results of Experiment 28 in Chapter 2 in this context. Would a sample of 2 drawing pins have been as good as 6? Would 100 be better than 6?

Sampling is treated at greater depth in many statistics books and the *quantitative* study of it involves quite complicated mathematics which is not at all suitable for our pupils. Teachers who want to study it for their own satisfaction will have no difficulty in getting suitable books. We have stressed the early *qualitative* ideas behind it, however, because they are within the grasp of some of the pupils and consideration of them is an important part of general knowledge about polls and surveys.

We stress in particular

(a) the need to match the sample and the population,

(b) the central idea that (subject to certain conditions being satisfied) the incidence in the sample is the same as the incidence in the population.

We hope to make it clear to pupils that a devious, or ignorant, operator can persuade the uninformed, or those who are over impressed by anything mathematical or scientific, to believe things which are not necessarily true. If we think about it we realize that even an expert and honest operator will very seldom be able to make a strictly accurate statement as a result of a sampling method – neither does he pretend to be able to do so.

Some experiments directed specifically towards discusssion of sampling.

Use the bottle which is half full of an equal number of red and yellow beads and take a sample of ten beads. Are five yellow and five red? Will they be for all/most of the samples you take? Take many samples and record the number of red and yellow beads. Repeat the experiment for a sample of seven beads. Then repeat if for samples of other sizes and compare the frequency graphs for samples of other sizes.

Take a sampling bottle which you cannot see through and ask someone else to fill it half full with beads of two colours in any proportion they like. Take samples to try to estimate the proportion of beads of each colour. When you have reached a conclusion check by counting all the beads.

Using the bottle of ball-bearings and beads, take samples to try to estimate the proportions of each. Check by counting afterwards. Do you think that there is any special bias in this experiment?

Take the bottle of red beads and add 50 yellow beads to it. Take samples to try to estimate the proportions of each colour in the bottle. On the basis of your answer how many red beads do you think there are in the bottle? Check by counting.

Take the bottle of blue beads and replace 50 of the blue ones by white ones. Do a sampling experiment to try to estimate the proportions of each colour. What would be your estimate of the total number of beads in the bottle? Check by counting.

9

Calculating Probabilities

Probability expressed as a vulgar fraction, decimal fraction or percentage – statistical probability – the addition law – the multiplication law – two non-independent events.

The idea of a measure of probability was first mentioned at the beginning of Chapter 1. Many of the activities 1 to 21 of Chapter 2 will have led to the idea that this measure concerns the comparison of the number of equally likely ways in which a particular event can arise and the total number of equally likely possible outcomes of trials. The comparison can be expressed as the fraction:

$$\frac{\text{Number of equally likely ways in which the event can occur}}{\text{Total number of equally likely possible outcomes of trials}}$$

This formula is applicable only to *a priori* situations (i.e. those with a finite number of outcomes). Consider, for example, the theoretical probability of scoring 7 with two fair cubical dice randomly thrown. The equally likely possible outcomes are the 36 ordered pairs listed on page 38 in Chapter 4. (Note here that the actual scores 2, 3, 4, ... 11, 12 are also possible outcomes but are *not equally likely* and hence inappropriate in this calculation. This idea was discussed in the analysis of Experiments 1, 7, 14 and 21 in Chapter 4.) Out of the 36 ordered pairs 6 of them, (1, 6): (2, 5): (3, 4): (4, 3): (5, 2): (6, 1) give a score of 7. We say, therefore, that the theoretical probability of scoring 7 is $\frac{6}{36}$. This value may be expressed in many ways (e.g. $\frac{1}{6}, \frac{2}{12}, \frac{3}{18}, \frac{12}{72}, \ldots\ldots\ldots 0{\cdot}1\dot{6}$ or $16\frac{2}{3}\%$). When probabilities are to be compared with each other the vulgar fraction notation is often inconvenient as it is sometimes difficult to appreciate the relative sizes immediately when two vulgar fractions have different denominators (e.g: which is larger, $\frac{6}{13}$ or $\frac{7}{15}$?) This difficulty vanishes if decimal or percentage notation is used (e.g: $\frac{6}{13} \approx 0{\cdot}46$ or 46% and $\frac{7}{15} \approx 0{\cdot}47$ or 47%).

If an event is an impossible one then obviously the number of ways in which it can occur is 0, so, the numerator being 0, the theoretical measure of probability is 0 also. If an event is certain (i.e. it will happen for all possible outcomes) then the numerator and denominator of the fraction are equal and hence the theoretical probability is 1. Thus every measure of probability lies between 0 and 1 or between 0% and 100% according to the notation used.

We hope that teachers will, by careful planning of experimental work and skilful questions, be able to let the pupils suggest all of the ideas mentioned so far in this chapter rather than, as some texts advocate, showing them how to calculate probability and then providing enough examples to make sure the technique is fully acquired. We suggest this course because of the aims which we stated on page 1. Naturally this does not preclude the situation in which the teacher may wish to ask pupils who have worked the experiments, drawn conclusions and discussed the ideas of this chapter, to calculate a variety of theoretical probabilities and we hope that pupils will do this with real understanding of the principles involved. It would be unreasonable to infer from our remarks that the demonstrate/practice method cannot lead to such understanding but we believe that the risk of its not doing so is greater than with the method based on practical experiment. If, however, our aims are not acceptable, and that of 'ability to compute theoretical probabilities rapidly in an abstract situation' is substituted, then there is evidence to suggest that the demonstrate/practice method will, in the same amount of time, more nearly achieve that aim.

We have not yet considered the method of calculating probability in situations which do not have a finite number of outcomes (i.e. Nos. 22–32 in Chapter 2). In discussion of this point pupils are likely to suggest, rightly, that a measure may be obtained from past or present statistics by the fraction

$$\frac{\text{Number of times the event occurred}}{\text{Total number of events}}$$

For example, in the group whose shoe sizes were examined in Experiment 25 (see Fig. 69) the probability that any boy chosen at random has a shoe size of $7\frac{1}{2}$ or over is given by

$$\frac{\text{Number of boys of shoe size } 7\frac{1}{2} \text{ or over}}{\text{Number of boys whose shoe sizes were recorded}}$$

In this case the result is $\frac{32}{55}$ or 58% approximately.

(Note that it would not be correct to say the probability was

$$\frac{\text{Number of sizes of } 7\frac{1}{2} \text{ or over}}{\text{Total number of sizes}} = \frac{6}{14}$$

as not all *sizes* were equally likely.)

We doubt that many teachers will wish to lead pupils for whom we write beyond this point on the topic of probability – some may not go so far. Nevertheless we will briefly mention some other ideas which occur quite early on in some texts (mostly with aims dissimilar to ours).

Suppose one has a red die and a blue die. The probability of scoring 5 if they are thrown together is $\frac{6}{36}$, the probability of scoring 4 is $\frac{3}{36}$. What is the probability of scoring *5 or 4?* Examination of the 36 ordered pairs shows that it is $\frac{9}{36}$ which is the *sum* of the first two fractions i.e. $p(5 \text{ or } 4) = p(5) + p(4)$ (using $p(5)$ as shorthand for 'the probability of scoring 5').

From a study of this and similar situations we begin to appreciate the so-called *addition law* which states that if a number of events are independent (i.e. they cannot happen simultaneously) then the probability that one of them will occur is the sum of the individual probabilities, i.e. if a, b, c . . . are independent events then

$$p(\text{a or b or c} \ldots) = p(\text{a}) + p(\text{b}) + p(\text{c}) + \ldots$$

Now consider the red and blue dice again. The probability that the red die shows 6 is $\frac{1}{6}$ and the probability that the blue die shows 6 is also $\frac{1}{6}$. What the probability is that *both* will show 6? Examination of the 36 ordered pairs gives it as $\frac{1}{36}$ i.e. the *product* of the two individual probabilities, so $p(\text{red 6 } \textit{and} \text{ blue 6}) = p(\text{red 6}) \times p(\text{blue 6})$.

After similar examples we arrive at the *multiplication law* that states that if a number of events are independent (in this context meaning that the occurrence of one does not influence the occurrence of the other) then the probability of them *all* occurring is the product of their individual probabilities.

i.e. $p(\text{a and b and c} \ldots) = p(\text{a}) \times p(\text{b}) \times p(\text{c}) \ldots$

We have stressed *independence* of events – a counter-example will show why. What is the probability that at least one 6 will be thrown by the two dice?

The two individual probabilities are both $\frac{1}{6}$. The ordered pairs show that the required probability is $\frac{11}{36}$ so clearly the addition law does not hold. This is because the two events can both happen at the same time and are not, therefore, independent in the sense of the addition law. Further examination shows that:

$$\frac{11}{36} = \left(\frac{1}{6} + \frac{1}{6}\right) - \frac{1}{36}$$

i.e. the probability of one or the other minus the probability of both. Such examples lead to a combination of the addition and multiplication laws for two events which can occur simultaneously, namely:

$$p(\text{at least one of a and b}) = p(\text{a}) + p(\text{b}) - p(\text{both a and b}).$$

We have gone far enough for our purpose. Those who wish to continue should consult the books in Appendix 1. In particular they may like to study the approach to probability through sets and to compare the general laws we have indicated with the laws appertaining to the number of members of mutually exclusive sets or of intersecting sets.

10

Comparing Statistics (1)

Percentages – statistical measures of central tendancy, mode, median, arithmetic mean, geometric mean, harmonic mean, 'batting' average.

The use of percentage notation to compare fractions has been mentioned and any collection of relevant newspaper cuttings made by pupils is sure to provide a great deal of material for a study of percentage methods.

One might introduce the idea by means of two general knowledge quizzes, one having 20 questions and the other 25. How can we compare scores in the two competitions? On what basis can we say performances are better, worse or the same? Some children will see no problem. They will say that $\frac{25}{25}$ is as good as $\frac{20}{20}$ and surely $\frac{24}{25}$ 'is the same as' $\frac{19}{20}$. What corresponds to $\frac{10}{20}$ or $\frac{4}{25}$ then? Suddenly it is not so simple and a lively class may have a number of ideas for dealing with this. One such, when the idea of proportionality has emerged, may well be to express all the marks as though out of 100 (hence the choice of 20 and 25 questions) and the idea can then be expanded to other totals, including those which are not divisors of 100 as 20 and 25 were.

Comparison of individual numbers, obtained by some sort of measure and/or calculation, is a relatively simple idea. Not so simple, but often necessary, is the comparison of two or more *collections* of data. Perhaps it is not surprising that in almost all cases the comparison is reduced to one of simple numbers again and it is the calculation of such numbers, known as *statistical measures*, which occupies a good deal of a statistician's time.

Consider the data collections obtained in Experiment 27 and discussed in Chapter 5. Fig. 72(e) shows a frequency graph of the heights of some boys. If you had to choose at a glance one interval as representative of the heights of the group you would probably choose the 165/170 cm interval because it is the most frequent. This is an acceptable statistical measure and and is called the 'mode' of the heights (mode – fashion). It is of limited use, being purely descriptive. Graphs (a), (b) and (c) in Fig. 72 show one of its limitations immediately. It is possible to have two or more modes in one set of data. Another problem with grouped data like this is that the size of the interval determines the mode – compare the actual modes in (a), (b) and (c). In graph (d) we get a much higher mode than in (e). Which seems the most satisfactory to you? Probably the one from (e) is your choice. If so how were you influenced? Perhaps by the fact that the mode has a more central position in (e) than in (d)?

The idea of central position gives rise to another much used statistical measure. If the boys whose heights were recorded were lined up in height order what would the height of the middle boy be? Graph (a) shows that it was 170 cm and this height is then said to be the 'median' of the collection. Unlike the mode it is unaltered by the grouping (except in so far as the grouping affects the accuracy with which it can be stated).

Another very common measure is based on the idea of 'fair shares for all'. We mentioned this in the discussion of Experiment 29 in Chapter 5. The girls compared the half-day absences for classes of different sizes by sharing the total absences equally amongst the number of pupils in the class – hence *pretending* that each member of the class had had the same number of half-days off. A similar suggestion would probably arise in relation to a discussion about pocket money. If we used this sharing idea what number would describe the heights of the boys previously discussed? Pupils may imagine a line of head-to-toe bodies across the playground being chopped into equal lengths! A feasible method, useful with a young class, is that of sticking paper strips together and finding the 'equal share length' which can then be compared with the group. This method is particularly useful if the group size is 8 (folding the strip into eight pieces is easy) and if the teacher then takes the place of one member of the group, the change caused by the substitution of one (presumably much taller) member can be seen.

The statistical measure found in the way outlined above is known as the 'arithmetic mean'. It is an example of an 'average' – it is, however, so frequently used that it is often called *the* average and many people are unaware that the term 'average' can be used for *measures of central tendency* other than the arithmetic mean. The median is also an 'average' under this definition and opinion is divided about whether or not the mode, which may not be *central* but is not really useful unless it is, can also be classed as an average. We would recommend the use of the specific names rather than the collective one.

These three measures, the mode, median and arithmetic mean are normally all discussed in modern programmes of school mathematics. It is unfortunate that sometimes 'average' is still taught as 'what you get if you add them all up and divide by the number there were'. To do this places all the emphasis on how you calculate *one* form of average (the arithmetic mean) rather than on what averages are and gives rise to the idea that the terms average and arithmetic mean are synonymous. Perhaps *the* average was introduced, when teachers began using *applications* of the four rules of arithmetic as well as the four rules themselves, for this was a valid example of the use of addition and division. Probably, at that stage, statistics was not studied in most schools in its own right (if it had been, one would have expected the simpler mode and median to come first).

Although the aim of practising calculations by means of applications would have been achieved, the backlash was that many pupils had their ability to accept a wider meaning of the word curtailed. Perhaps such considerations should make us think more carefully about applications for the mathematics we teach lest in attempting to provide motivation for practising techniques we obscure future work.

It is also worth noting that whilst the modal value always appears in the population, the median and the arithmetic mean may not (at one time the average number of children per family was 2·8. Notice too that the mode is easily seen from a frequency graph and the median is equally easily obtained from a *cumulative frequency graph* (see Experiment 20, Fig. 61).

Fig. 102 below shows such a graph of the shoe sizes of the boys measured in the analysis of Experiment 25.

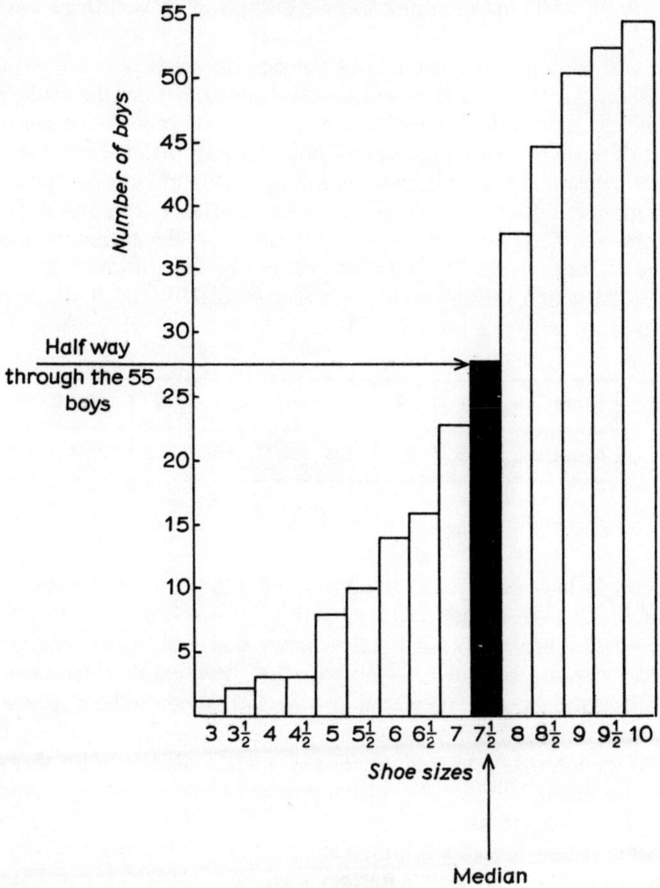

Fig. 102

If children are encouraged to bring newspaper cuttings which mention averages, or containers on which 'average contents' or 'average life' are stated, it is worthwhile discussing just which averages are thought to have been used and how they might have been obtained. Does the match manufacturer add up all the matches and divide by the number of boxes? Does the 'average wage' of workers include the low-paid women and the highly paid managers?

It is often difficult (but not impossible, especially when wage claims are news) to find real examples in newspapers or advertisements where a more appropriate average might have been chosen. Sometimes a hypothetical example can help to pinpoint the idea. E.g. if nine paper-boys earn £1 a week and one other earns £2, would the arithmetic mean, the median or the mode be the 'best' average to use to describe the average earnings of those boys? ('Best' will of course initiate discussion of 'best from whose point of view?')

The arithmetic mean is often just known as *the mean*, but this again is not a useful practice because there are other means besides the arithmetic one. One example is the '*geometric mean*', which we include for the interest of teachers. A simple example shows how it differs from the arithmetic mean. It is used in cases of multiplicative growth. Suppose a sum of money is doubling each year and is £16 after 4 years and £1024 after 10 years. What was the sum after 7 years? 7 is the arithmetic mean of 4 and 10, but obviously the amount cannot be the arithmetic mean of 16 and 1024 which is 520 for this would give £1040 after 8 years. A table shows that the answer required is £128.

Years	4	5	6	7	8	9	10
Amount	16	32	64	128	256	512	1024

Fig. 103

This might have been found my multiplying the numbers 16 and 1024 and taking the square root, i.e. $\sqrt{16 \times 1024} = 128$. This example is of course trivial because, knowing the money was doubled we only need the 4-year amount to be able to compute the 7-year amount. However, in a case where the growth rate was unknown (but known to be a steady multiplicative one), this new mean, the geometric, would be the way to compute the intermediate amount. The geometric mean is often the appropriate one in situations involving growth of population, or of money at compound interest.

Another situation uses a third named mean. If 100 metres is covered at 5 metres/second and the return journey made at 2 metres/second what is the 'average' speed – i.e. the constant speed which would have covered the same total distance in the same time? Is it 3·5 metres/second – the arithmetic mean of 2 and 5? The total distance is 200 metres and the total time taken is $(\frac{100}{5} + \frac{100}{2})$ seconds hence the constant speed would be

$$\frac{200}{100\,(\frac{1}{5} + \frac{1}{2})} \text{ or } \frac{2}{\frac{1}{5} + \frac{1}{2}}$$

(just under 3 metres/second). The number is known as the 'harmonic mean' of 2 and 5 and the expression $\frac{2}{\frac{1}{5} + \frac{1}{2}}$ shows how it is calculated from the original figures. The 2 and the 5 of the original speeds appear in the denominator and the 2 in the numerator appears as we are dealing with the harmonic of *two* numbers. This mean is often the appropriate one in questions of speeds, rates and prices.

We have discussed three types of mean. In summary they are calculated as follows:

Mean	Of 2 numbers	Of n numbers
Arithmetic . .	$\frac{a + b}{2}$	$\frac{a + b + c + d + \ldots}{n}$
Geometric . .	$\sqrt{a \times b}$	$\sqrt[n]{a \times b \times c \times d \times \ldots}$
Harmonic . .	$\frac{2}{\frac{1}{a} + \frac{1}{b}}$	$\frac{n}{\frac{1}{a} + \frac{1}{b} + \frac{1}{c} + \frac{1}{d} \ldots}$

Fig. 104

Our purpose in discussing the means is to show that there is a greater variety of useful statistical measures than many people realize. Several of the books in Appendix 1 study them in greater detail. Let us resist the temptation to use problems on geometric and harmonic means solely in order to practice roots and reciprocals! We feel, however, that the ideas behind the use of these numbers, as opposed to their repeated calculation, may be of interest in some situations.

We conclude this chapter with a look at an average with which many boys will be familiar – the cricketer's batting average. It is very like an arithmetic mean, but the total number of runs is shared not amongst the number of 'innings' but amongst the number of 'outings'. Suppose a batsman scores a total of 100 runs in 6 innings, but is 'not out' in one of them then the batting average is said to be 20 i.e. $\frac{100}{5}$, 5 being the number of times 'out'. What would the average be if he was 'not out' in all of the 6 innings? Certainly not $10\dot{0}$ as that would be the average for 5 'not outs'.

11

Comparing Statistics (2)

Statistical measures of dispersion (range, quantiles, mean and standard deviations) – Special theoretical distributions (Binomial, Normal, Poisson) – Correlation.

So far our statistical measures have been designed to allow us to substitute one number for a set of numbers. However, it is possible for two *different* distributions of data to have the same mode, median, arithmetic mean, or any combination of those three. Such is the case for the two distributions typified in Fig. 105 – notice that in this example both are symmetrical. Is this a necessary condition for them to have the same mode, median and arithmetic mean?

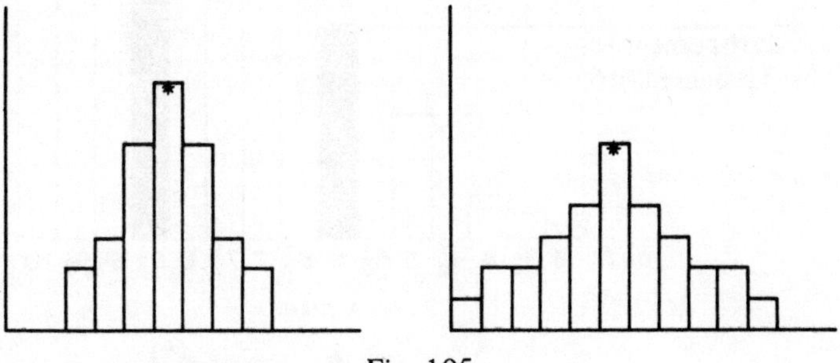

Fig. 105

How would you describe the differences between the two distributions?

One way is to state the lowest and highest measures in each case – in other words to describe the *range* of results. This measure does not use all the data as it is specified by only two of the readings and it often happens that the extreme values in a distribution are, for one reason or another, very much 'out on their own'. For this reason a limited range is often quoted instead, to eliminate extreme cases. The most common part-range is linked to the idea of the median. The data are arranged in ascending order and then split into 4 equal parts, then the range from the end of the first quarter to the beginning of the last quarter is taken. The two end points are called the 25th and 75th percentiles or the 1st and 3rd quartiles, hence the name given is the 'interquartile range'. This method shows up well on a cumulative frequency graph, as shown in Fig. 106.

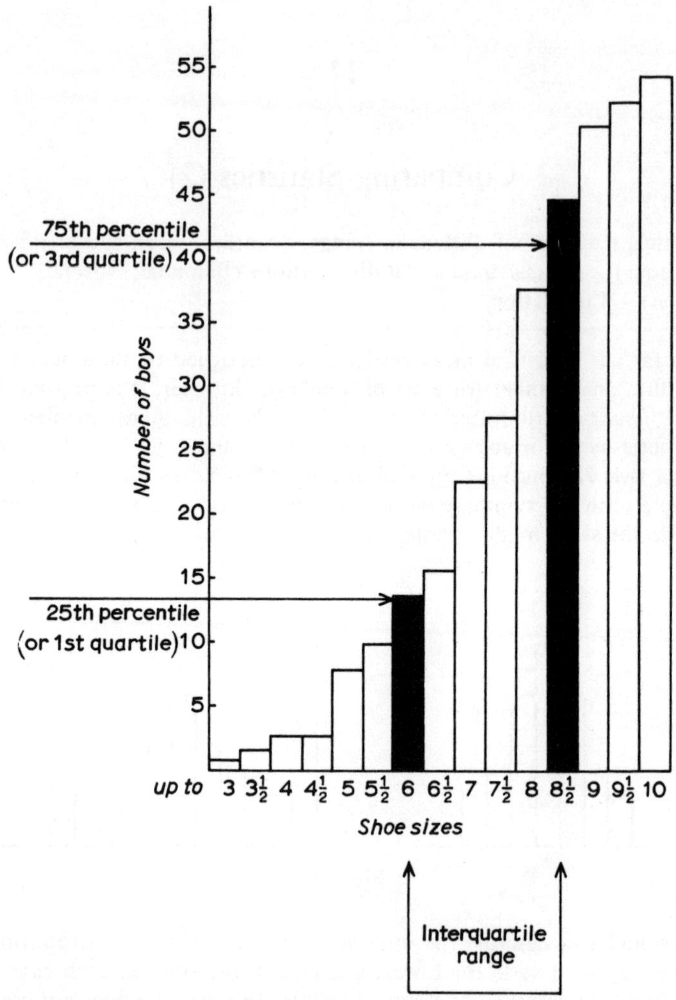

Fig. 106

Such a division is not always into 4 parts. Any number will do and a division into ten parts, each called a *decile*, is often favoured. This general method, of ordering and dividing the data into equal parts, is known as taking *quantiles* (not to be confused with quartile which refers only to 4 parts).

The range and inter-quantile ranges are the simplest statistical measures of *dispersion.* Two other such measures, which need considerable calculation, but which teachers may have met in their own training, are the *mean deviation* and the *standard deviation.* We do not recommend them for the pupils we are considering, but teachers may like to find out more about them from a book on statistical methods.

During the experimental work and in our consideration of statistical measures we have made a great deal of use of the frequency or counting graph. In particular we have often come across the type of frequency distribution which is 'heaped up in the middle'. Such a graph arises in many cases where classification of the population is on the basis of some measurement (e.g. physical, mental, performance, . . .). It is one of the most striking appearances of pattern in mathematics and we hope that it will be observed by pupils during the discussion and comparison of results.

Pattern so often arises in statistical data that there are a number of named *theoretical* distributions which, under certain circumstances, are assumed to 'fit' particular types of data. Although our pupils will be expected to do no more than notice similarities and differences between distributions obtained from different sources, teachers may themselves wish to study the distributions and their applications in detail and they are well expounded in several of the books in Appendix 1. One such is the *binomial distribution* which is applicable in situations involving successive discrete events such as occurred in many of the coin-tossing and dice-throwing experiments. Not surprisingly, then, it is related to Pascal's Triangle and, more generally, to the Binomial Theorem in algebra.

Another theoretical distribution is the so-called *Normal Distribution* (known as 'bell-shaped') which is often applicable where measurement on a continuous scale is involved, e.g. heights of a certain age group, intelligence distribution . . . A third type, is the *Poisson Distribution* which might have had application to the football data and children in families data collected earlier.

We have paid considerable attention to the comparison of data involving classification of the population and giving rise to frequency distributions. We must also consider briefly the ways of measuring and comparing the relationship between two sets of data. In Chapter 3 we discussed relation graphs and Experiment 27 (see Chapter 5) gave the data for a good example of a graph of this type.

If it so happened that height and armspan were in direct proportion then the points on the graph would be on a straight line thus:

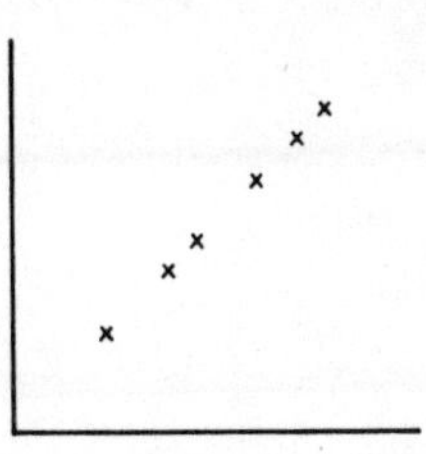

Fig. 107

If they had no relationship to each other at all then one would expect an overall scatter thus:

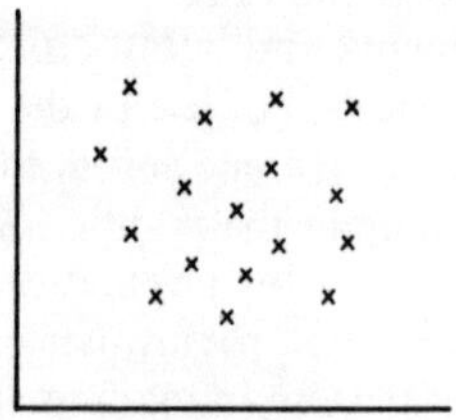

Fig. 108

There are a variety of ways of assigning a measure to the amount of *correlation* between two sets of data. Again we recommend teachers to find some of them in a book on statistical methods. In all cases the perfect direct relation is given the number +1 and to the perfect inverse relation is assigned −1. A value of 0 represents no correlation and other amounts lie between the two extremes.

There is one very big pitfall however. Such a number, known as a correlation coefficient, can be assigned to any two sets of data which can be represented on a relation graph. In cases where the coefficient is calculated as 0 we may say that there is no causal connection between the two sets of data. It frequently happens, however, that two sets of data *cannot* be said to have a causal connection and yet they have *a non-zero correlation coefficient.* For example, a high positive correlation coefficient arises from a comparison of the yearly numbers of motor accidents and packets of cigarettes purchased. Does this mean that smoking causes road accidents (or vice-versa)? Probably it does not. Statistics does not offer justification for assigning any causal meaning to a correlation coefficient. Other independent evidence (often biological or social in origin) would be needed to point to such a link between the two sets of data. Misuse of the ideas of correlation, like misuse of sampling, contributes to the scepticism with which some people regard statistical methods.

12

Suggestions for Follow-up Work

Several times already we have suggested further work which could be undertaken as a result of interest kindled by one of the activities we have described. We could now finish by citing some particular applications of ideas we have considered. We do *not* do so because to choose just a few from the many may make them look more important in their own right than they are.

Pupils who have worked this far will probably now be in their last year at school, and it is more appropriate than ever that they should be working independently in school and having the opportunity to visit establishments and talk to people outside school. For this reason we suggest, in this final chapter, some possibilities for projects. The source material for them comes from local industry, administration, commerce, libraries, people . . . and any other means that pupils and teachers can find.

1. Try to visit a variety of local establishments covering scientific, commercial, engineering, social, administrative, advertising, insurance interests to find out what use is made of statistics and probability. Look in particular for market research, record keeping, stock and quality control, scientific results, population statistics, random sampling. It is helpful if both sides know what particular things pupils have studied or have been asked to notice.
2. How do advertisers, the Press and T.V. use statistics and make predictions? Collect examples of graphs and interpretations of data from papers and T.V. Study the *misuse* of graphical methods to produce certain effects on the reader or viewer (this topic is discussed very well in several of thc pupils' books suggested in the booklist).
3. The what, why and how of the Retail Price Index.
4. The mathematical theory of probability is relatively modern – mostly less than 300 years old. A number of interesting historical pathways to it could be explored using a variety of books as sources of information.
5. How are the ideas we have studied used by biologists, physicists and chemists? This presents an opportunity to bring teachers of other subjects into the work.
6. Do you think that the ways in which examination results are compared and/or combined in your school are statistically sound?

Appendix 1

A list of some books which contain material related to the work of this guide

A. Topic books for pupils' classroom use.

B. Books for pupils' or teachers' reference as part of a classroom library.

C. Books for teachers rather than pupils.

We suggest that copies of the books listed might be available, for teachers' inspection, at Teachers' Centres.

A. Topic books for classroom use suitable for many average and below average 13–16 year olds.

Probability	Razzell and Watts	(Hart Davis)
Starting Statistics	Lewis and Ward	(Longmans)
Statistics	Campbell	(Longmans)
Computers	Fielker	(Cambridge)
Statistics	Fielker	(Cambridge)
Towards Probability	Fielker	(Cambridge)
The World of Statistics	Johnson and Glenn	(Murray)

B. Books for reference by pupils or teachers, suitable for a classroom library

Pathways to Probability	King and Read	(Holt)
Mathematics in the Making	Hogben	(MacDonald)
The World of Mathematics	Kasner and Newman	(George Allen and Unwin)
A Short History of Mathematics	Rouse Ball	(Dover)
Mathematics in Management	Battersby	(Pelican)
How to lie with Statistics	Huff	(Gollancz)
Annual Abstract of Statistics		(H.M. Stationery Office)
The Guinness Book of Records		(Doulton)
Men of Mathematics	Bell	(Pelican)
How to take a chance	Huff	(Pelican)
Mathematics		(Life Science Library)

Learning with colour. Mathematics	Adler	(Hamlyn)
Which? In Secondary Schools (Mathematics and Science)		(Consumers' Association)
The Language of Mathematics	Land	(Murray)
Mathematics and the Imagination	Kasner and Newman	(Pelican)
Probability and Statistics	Sherlock	(Arnold)
How to read statistics	Vesselo	(Harrap)

Which? The monthly bulletin of the Consumers' Association. (Classroom copies available at reduced rate.)

C. Books more suitable for teachers than pupils

Probability and Statistics	Nuffield Mathematics Project	(Chambers/ Murray)
Probability and Statistics for Everyman	Adler	(Dobson)
Introduction to Finite Mathematics	Kemeny, Snell and Thompson	(Prentice-Hall)
The Use and Abuse of Statistics	Reichmann	(Pelican)
Facts from Figures	Moroney	(Pelican)
Sets and Logic 2	Bailey	(Arnold)
Statistics – a first course	Loveday	(Cambridge)
Statistics – a second course	Loveday	(Cambridge)
The Scope of Mathematics	Holt and Mackintosh	(Oxford)
Statistics and Probability	Glasgow School Statistics Panel	(Foulsham)
Chance and Choice	D'Arcy	(Thames and Hudson)
Probability and Statistics	Ormell	(Oliver and Boyd)
Breakthroughs in Mathematics	Wolff	(Signet Science Library)
Statistics: the How and Why	Lockwood	(Murray)
Chance and Choice by Cardpack and Chessboard	Hogben	(Parrish)
Lady Luck, the Theory of Probability	Weaver	(Heinemann)
Educational Statistics for Beginners	Griffiths and Downes	(Methuen)

Appendix 2

Presentation of Instructions

In the introductory paragraphs of Chapter 2 we made some remarks about ways in which instructions could be given to pupils. We have not attempted to produce a set of assignment cards or worksheets on this topic, for teachers to hand out to their classes, because whatever we were to produce would suit only a small minority of classes.

There are many variables which affect the amount of detailed instruction which must be given to pupils doing practical work. Such features include the pupil's ability to read and comprehend and his ability to make decisions without consulting the teacher at every step; this last is dependent in part on personality and past experience in the classroom. One must also consider one's own organising ability, with respect to both pupils and equipment, and one's readiness to cope with spontaneous questions and points needing discussion whilst moving round the class.

There are many other factors too and together they must have an effect on the written and spoken communication between teacher and pupil, especially where these concern instructions for procedure. Therefore we cannot offer a universal solution to the communication problem but we give one or two examples of current practice. Sometimes teachers who are inexperienced in organising practical and/or individual work make things difficult for themselves. For example, imagine a teacher taking a new class, whose members are not used to practical work, trying to get a wide variety of activities going by handing out workcards copied from Chapter 2 of this book. We should not be surprised if the result were chaos and the teacher were to decide that practical work is just not possible with this class. Most children do not have a natural ability to make their own decisions about procedure, recording, etc. but they can learn (at varying rates) to do so. If the pupils do not know the teacher and his ways of working, or if they themselves are not used to practical or individual work, or if the language of the communication is too difficult, then the number of questions will be high and the teacher may not be able to cope with them all adequately.

Such a situation could be improved by making sure that the language was right for the pupils; by giving (at first) enough instructions to avoid too many questions about procedure; by limiting the number of different activities and by staggering the start of practical work by different groups of children so that only some of the class are actually beginning a new experiment in any one lesson. Of course it is only with experience that one learns how long particular activities are likely to take. It is really unnecessary for us to say that the situation is further simplified if pupils know where things are kept and if they are trained to put equipment, books and paper away after use.

In contrast, the brevity, which we *tried* to achieve, in the suggestions of Chapter 2 might well be appropriate for a class of pupils of average reading ability, already used to an assignment system. In such a case the teacher might simply add our suggestions, where appropriate, to the existing system.

Now we give a typical example of a workcard for insertion in an existing assignment system; an example of a worksheet which might be used with pupils just beginning practical work; and an example of a non-written communication. In each case the work concerns Experiment 8 of Chapter 2, hence the mathematics involved is the same.

(a) A workcard for an existing assignment system.

Experiment 8

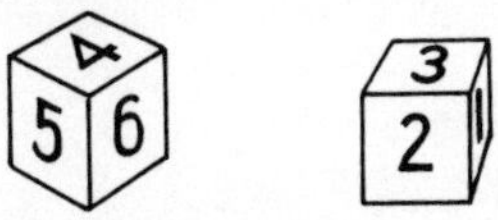

If you throw two dice and add the scores what is the highest possible total?

What is the lowest possible total?

What other totals could you have?

NOW THROW TWO DICE 150 TIMES AND RECORD THE TOTAL OF EACH THROW.

Try to explain why some scores come up frequently and others hardly at all.

Such a card presupposes that the pupil knows how to record results in a systematic way and hence will be unlikely to need to ask the teacher for further instruction. It also implies that he is used to looking for pattern and used to looking for reasons behind what he observes. The card is, therefore, more likely to be worked successfully if the pupil has been well trained and and feels secure in his relationship with the work and with the teacher.

(b) A duplicated worksheet for use by individual pupils who need more detailed instructions.

It is likely that completion of such a worksheet would take pupils about $1-1\frac{1}{2}$ hours during which time they should not need much individual attention from the teacher, provided that the language level is right. The major teacher/pupil contact comes when the worksheet is complete.

We would hope that pupils who need to start with this amount of detail would gradually move to the sort of communication suggested in (a). The mathematics learnt in both cases should be the same, but the former adds to the pupil's development as a person who can organise his own work and make some decisions for himself.

Worksheet No. 8 *Name*......................

You will need 2 dice, one red and one green

Throw the dice together and add the two scores.
Write down your total score. ....
What is the smallest score it is possible to get? ....
What is the largest possible score? ....
Now throw the 2 dice at least 150 times and record each score by a small mark in the columns below.

SCORE	2	3	4	5	6	7	8	9	10	11	12
Number of marks in each column											

Look at the number of times each score has been thrown.
The lowest and highest scores have not been thrown very often, but the scores in the middle appear quite a lot. Try to think of a reason why this nearly always happens and write it here.

....

....

....

Now turn over the page.

There is only one way of throwing a score of 2, it is

Red	Green
1	1

There are two ways of scoring 3. Fill them in.

Red Green   Red Green

Fill in the ways of scoring 4.

Red Green   Red Green   Red Green

Fill in the rest of the scores.

5   6   7   8   9   10

11   12

Use what you have just done to try to explain why 7 is the most likely score to throw with two dice.

....

....

....

(c) It often happens that it is necessary to give fairly detailed instructions to children whose ability to read, and to comprehend the written word, is severely limited. In such cases (and indeed in others too) some teachers have been able to make good use of a tape-recorder. The tape can be replayed as required. Cassette tapes are particularly useful as they are very simple in operation and can be used on a play-back only machine. Such machines are not expensive and only one actual cassette *recorder* is needed in the school. Individual headphones can be used to cut out noise. We give below the script of a tape which could be used in this two dice experiment. Sometimes it is more convenient to tape a worksheet so that pupils have the written and spoken word together.

This tape is for Experiment No. 8. It is an experiment with two dice, one red and one yellow. As well as the dice you will need a shaker and tray, a sheet of squared paper and a pencil or pen. Switch off and fetch them if you need to.

Shake up your two dice and throw them into the tray. Say what you scored on the red die . . . . Say what you scored on the yellow die . . . . Now add the two scores together and say your total . . . .

I wonder what it was? I shouldn't be surprised if it was 6 or 7 or 8. What is the highest score you could possibly throw with two dice? . . . . Yes, did you say twelve? You would get twelve if you threw two sixes. What is the lowest score you could get with two dice? . . . . It would be a score of 2, that would be a 1 on each die.

Now take your squared paper and write the numbers from 2 up to 12, one to a square, across the top of the paper. Switch off while you do it.

Let's check. You should have written the numbers 2, 3, 4, 5, 6, 7, 8, 9, 10, 11, 12 in the squares across the top of the paper.

Now throw the dice again and add up the scores. Find this number at the top of your paper and fill up the square underneath it with a cross.

Shake up the dice and throw again. What have you scored this time? . . . . Put a cross under that number on the paper . . . . I wonder if you got the same score both times. Probably not but if you did you will have two crosses one under the other.

Now I want you to go on by yourself throwing the dice and marking the the total score by a cross on the paper. Go on until the column of squares under one of the numbers is full up. Switch off while you do this and afterwards come back to hear some more of the tape.

Well, I wonder which column was filled up first. I wouldn't be a bit surprised if it was 7. If not I bet it was 6 or 8. How many times did you score 2? . . . . How many times did you score 12? . . . . Can you think why it is that 2 and 12 didn't come up very often but probably 6, 7 and 8 came up a lot?

The next part of the tape will help you see if you are right. How many ways can you score 2? . . . . yes, just one way. The red die must show 1 and the yellow die must show 1.

How can you throw a score of 3? . . . . There are two ways this time, not just one. You could get a 1 on the red die and a 2 on the yellow die or the other way round, a 2 on the red die and a 1 on the yellow.

Turn your paper over and write down all the ways you could score 4. Switch off while you do it.

Did you get three ways? . . . They are red 3 and yellow 1 or the other way round, red 1 and yellow 3. Then there is another way because you could have a red 2 and a yellow 2.

In a moment I want you to switch off and write down all the ways you can score 5, then 6, then 7 and so on right up to the score of 12. After that come back to hear the rest of the tape. Now switch off.

Now we will check your answers.

First you wrote the ways of scoring 5. Did you get (red 4, yellow 1) and (red 1, yellow 4), (red 3, yellow 2) and (red 2, yellow 3)? That is four ways altogether.

I wonder if you noticed a pattern in the number of ways . . . . First it was one way, then two ways, then three ways and then in that last one there were four ways. If the pattern goes on how many ways would it give next time? . . . .

Let's check. The score is a total of 6 and there are five ways to get it. They are (r5, y1), (r1, y5), (r4, y2), (r2, y4) and (r3, y3). Did you get all of them? . . . .

Now for a score of 7: there are six ways this time. (r6, y1), (r1, y6), (r5, y2), (r2, y5), (r4, y3), (r3, y4).

How many ways did you find for a score of 8? . . . . Well there are only five ways, (r6, y2), (r2, y6), (r5, y3), (r3, y5) and (r4, y4). If you got too many ways I expect you forgot that you cannot throw higher than 6 on one die.

There are four ways to score 9. They are (r6, y3), (r3, y6), (r4, y5), (r5, y4).

For a score of ten there are only three ways, (r6, y4), (r4, y6) and (r5, y5).

As you probably expected there are just two ways for a score of 11. They are (r6, y5) and (r5, y6).

And lastly to score 12 there is only one way: both dice must show a 6.

This is nearly the end of the tape. When it is finished I want you to use the rest of the space on your paper (or take a new piece if you need it) to try to write down why you think a score of 2 and a score of 12 do not appear very often and why a score of 7 usually happens most in this experiment. Then give your paper to me. Don't forget to put the two dice away will you. That's all thank you. Switch off now please.

Appendix 3

Pupils' Written Work

Some teachers who are used to daily marking of exercises express concern about the change in the pattern of written work which must follow from the methods we suggest in this book. One cannot assume that pupils know automatically how to write up results of experiments even though some of them will have experienced this sort of recording in science.

We would suggest that, normally, it is unnecessary for pupils to spend time in writing an account of how an experiment was performed as the teacher will be already aware of this. The following points may help to give some indication of what one might eventually look for in presentation from pupils who have become used to this method of working. It must be borne in mind that unless pupils are given specific help in moving towards the expected standard then usually one will look for it in vain.

(a) Recording of raw data in a way which shows some forethought. (E.g. in a well thought out tabulation rather than in a random way 'as it comes'.)
(b) Some classification and/or ordering of data where appropriate, possibly illustrated by a graph.
(c) Some attempt by the pupil to record what he has noticed, or been helped to notice, in the data and his own efforts at explanation (see also (e)).
(d) Use of data for prediction of future results, where appropriate.
(e) Additions to (c) and (d) after follow-up analysis and discussion of results with the teacher and other pupils.

The marking of such written work usually takes longer per page than the marking of exercises, but the time factor resolves itself in the long run as pupils submit their results at different times and less frequently. Teachers who are not used to making subjective value judgements in their marking soon get used to it once they have their aims clearly in mind.

Course-work produced in this way is rightly influenced by teacher/pupil discussions during the work. If a teacher finds it necessary to examine his pupil's unaided ability then an assignment can be worked and recorded without the teacher's help. Such an assignment would usually be related to previous ones, not merely a repetition, so that the pupil has an opportunity to use his previous experience in a new situation. The marking of such examination assignments does not differ from the marking of classwork as the same ideals are applicable.